Larry - It is always a pleasure to sign this on behalf of Eileen, and her niece, Julia Bolton Holloway.

Karen

LICHENS FOR VEGETABLE DYEING

EILEEN M. BOLTON

SECOND EDITION

EDITED BY KAREN LEIGH CASSELMAN AND

JULIA BOLTON HOLLOWAY

ROBIN & RUSS HANDWEAVERS
McMinnville, Oregon
1991

© Copyright 1960 Eileen Mary Bolton
© Copyright 1982 Julia Bolton Holloway
First published by Studio Books, Longacre Press Ltd.
Reprinted in Great Britain 1972 by Studio Vista Publishers
Published in United States 1972 by Robin & Russ, Handweavers

Republished in 1991 by Robin & Russ, Handweavers,
533 North Adams Street, McMinnville, Oregon 97128

All rights reserved. No part of this publication
may be reproduced, in any form or by any means,
without permission from the Publisher

ISBN 1-56659-001-9

CONTENTS

Editors' Preface   4
Author's Preface   7

Chapter

   I   The Lichen Dyes   9
   II   The Lichen Plant   12
   III   The Orchil-Producing Lichens   15
   IV   The Boiling-Water Lichens   19
        Less Plentiful Lichens which give Dyes   27
   V   Extracting the Orchil Dyes   28
   VI   Extracting the Dyes, Boiling Water Method   32

Appendices on Acids   35
List of Reclassified (Renamed) Lichens   39
Bibliography   42
Index   43

EDITORS' PREFACE

This handbook for weavers, published originally in 1960 and 1972 in England and America by Eileen Mary Bolton, is now being reissued, due to Karen Leigh Casselman's quest first for the author, then for the owner of the copyright. Karen Leigh Casselman is a Research Associate in Natural History at the Nova Scotia Museum, Halifax. She is the author of *Craft of the Dyer: Colour from Plants and Lichens of the Northeast*, 1980; *Lichens and Dyes: An Historical Sourcebook*, 1991; and numerous articles. Julia Bolton Holloway, Director of Medieval Studies at the University of Colorado, Boulder, teaches students to appreciate the paradox of the delicacy, yet intensity, of the palette of colours used prior to aniline dyes.

It was when viewing herbarium specimens in Ottawa, Washington, London and Dublin that Karen Leigh Casselman realized the accuracy of Eileen Bolton's statement that a dried specimen is "a poor shrivelled ghost of its former self." In painting lichens in their correct relation to one another as regards size, colour and texture, Eileen Bolton returned to the seventeenth century tradition of botanical illustration characterized by a unique blending of art, craft, science and whimsy. She was the first modern craft book author to understand the role of lichen acids, and to seek expert advice from two continents. Jack Laundon, then with the British Museum (Natural History), London, and a past president of the British Lichen Society, challenged the Casselmans to find Eileen Bolton, when he found a Welsh address in the membership files. However, they came to Betwys-y-coed, North Wales, just after Eileen Bolton's death in August, 1981. In 1985, Maura Scannell, a botanist at the National Botanic Gardens, Dublin, showed Karen the dyed fleece samples that comprise PLATE VI. In Wicklow, Ireland, Karen met Eileen Bolton's brother and sister-in-law, the Very Reverend Frederick Rothwell and Mrs. Bolton; they gave her Julia Bolton Holloway's address in the United States.

Eileen Mary Bolton was the second of four children born to artists, John Nunn Bolton (see *Dictionary of Irish Artists*, p. 70-71) and Florence Bolton, née Francis, in Warwick, England. Eileen Bolton's father died when she was six. Related to Samuel Beckett, he had painted in water colours and oils, had won a gold medal in Dublin, Ireland, and had almost been elected to the Royal Academy in England. She wanted to attend Slade Art School but her mother, having to educate three other children, supporting them by doing embroidery for the Countess of Warwick and the Church of England, forbade her. So Eileen Bolton educated herself, refusing to work, except for a World War II period of teaching art to schoolchildren. She continued a lifelong quest for knowledge, doing woodblocks and stained glass windows, and researching the astronomy in Irish and Welsh poetry. She often dyed, wove and sewed her own clothes (using deep, soft blues and greens for these). Eileen lived first in Warwick with her sister, Dorothy Joyce Bolton, next did housekeeping for her brother, the Cambridge-educated Frederick Rothwell Bolton (author of *The Caroline Church in Ireland* and later a Dean of the Church of Ireland), in Nottinghamshire, and then settled in a National Trust weaver's cottage, Pont Eidda, Betwys-y-Coed, Wales. Her oldest brother, Oxford-educated John Robert Glorney Bolton, became a journalist and writer, publishing *The Tragedy of Gandhi, Peasant and Prince, Two Lives Converge, Pétain, Sir Christopher Wren, The Dome of Devotion, Czech Tragedy, Living Peter* and *Roman Century*. Eileen's younger sister, Dorothy Joyce Bolton, came to the United States at the outbreak of World War II. She taught at university level, authored *To Discover, To Delight*, and was a painter with many exhibitions.

Julia and Richard Rothwell Bolton, Glorney Bolton's children, went to stay with Frederick and Eileen Bolton in Nottinghamshire during World War II. Their uncle tutored them in their formal lessons while Eileen taught them to launder their clothes, to sew and to paint, had them herd her fierce geese and collect eggs from the henhouse. She took them on long walks to reach Sherwood Forest - we never did - and read them stories which they then acted out as plays, their audience being usually her dog Tavy. She had twinkling, mischievous grey eyes and was almost the children's size. Richard later visited her in Wales when he was a schoolboy at nearby Shrewsbury and they spent nights discussing ancient astronomy, including that of Stonehenge.

Eileen visited Joyce and Julia Bolton in America after a 1953 visit to Egypt. She told us about the stained glass window she had made of the Four Horsemen of the Apocalypse with the portraits of Hitler, Mussolini and Goebbels - which was destroyed in war bombing. She showed Joyce and Julia Bolton the black and white scrapper board illustrations she had done of Egyptians in villages, dashing to catch buses, in the markets, in cities, which were to have been published. Her wedding present to Julia was a corduroy hand-spun, dyed and woven carpet in soft purples, reds, golds and browns with an aroma like Harris tweed, which was treasured for a quarter of a century. Her empirical researches in lichen dyeing were mainly carried out in the British Isles, but she experimented in California as well.

Julia and her sons, Richard, Colin and Jonathan Holloway, travelling from America, visited Eileen in her Welsh cottage in 1972. The weaver's cottage is tiny in its proportions. It was the first house that Eileen felt was right for her. Just beyond its dimunitive lawn is a trout stream with a bridge where Eileen and the children gathered lichens. During the days she would take the boys to collect wool from the hedges, and then taught them how to dye the fleece shades of purple, brown, red and blue. She showed them how to tell the time on the sundial we had given her and at night played her Irish harp to them. We went off on excursions to Conway Castle and Plas Mawr. A brilliant, self-educated scholar, Eileen Bolton was also a gifted teacher.

On the walls of her small cottage hung examples of her mother's embroidery. In a second building were her spinning wheel and looms. Tucked away in drawers were books on Celtic mythology, astronomy and literature. On the Welsh dresser were delicate cups and saucers. When Eileen Bolton died at the age of 78, these treasures, along with her Irish harp, were lost. Her orbituary was published in the village newspaper in Welsh. This book, some articles on Celtic literature and astronomy, a woodcut of the Shakespeare Herd in Warwickshire, another of duffle-coated Welsh shepherds coming to the Christ Child, a scrapper board sketch of the goddess Isis, a painting of a Norwegian fjord, and some further sketches of lichens, each done in her meticulous style, survive. Maria Antonia Bandres y Bolton (Richard Bolton's widow in Rome), owns three oil portraits of Eileen as a child painted by her father, John Nunn Bolton, as well as landscapes painted by Eileen Bolton.

It is a tribute to Eileen Bolton's autodidactic scholarship that lichenologists at botanical institutes, museums and universities have *Lichens for Vegetable Dyeing* on their book shelves and cite it in their bibliographies. This book has provided the base for much, if not all, of the lichen dye research in Britain, Ireland, Canada, the United States, Australia and New Zealand during the last thirty years. Its reputation as a classic in international craft literature is justly deserved.

The editors wished to make *Lichens for Vegetable Dying* available once more to benefit a new generation of dyers, spinners and weavers. The integrity of Eileen Bolton's text has not been altered. Julia Bolton Holloway has described Eileen Bolton's life, and edited and typeset the text. Karen Leigh Casselman has provided synonyms and nomenclature, the status of rare and potentially endangered lichens, and an additional bibliography.

This second edition of *Lichens for Vegetable Dyeing* is dedicated to the memory of Eileen Mary Bolton.

Halifax, Nova Scotia
Boulder, Colorado
1991

AUTHOR'S PREFACE

This book has been prepared with the idea of helping the handweaver find and identify the dye lichens more easily, and to assist in the making of dyes from these plants. The drawings were made from the lichens in their natural surroundings, and the plants have been grouped together, so that their relative sizes, colours and textures may be compared. The Welsh topsnails, bracken-shoots, butterflies, etc, have been included to give scale to the plants, for they are mostly rather small. Many distinguishing features, too small for the eye to see, have been enlarged on a separate Details Plate, together with any spores which have been observed.

I would have appreciated this handbook when I first set out to dye lichens. I wasted much time by not having a clear idea of what to look for in the first place, particularly with regard to their sizes and colouring when moist, which is when they are gathered. The most exact botanical description fails to create a living image in the mind, and a dried specimen is a poor, shirvelled ghost of its former self. However, when a few possible plants are in the hand, a good botanical description is essential for their identification. How then can the handweaver, who may not be a botanist, avoid the arduous task of mastering a glossary on the lichen plants, before beginning to understand the technical literature on these plants? Likewise, a series of drawings, all of one size, whether they be of a part or the whole of a plant, can be very confusing, even if the magnification or reduction is stated alongside.

In order to spare the handweaver much lengthy study, and to provide a simpler handbook for such use, I decided to draw the lichens in their correct relation to each other as regards size, colour and texture. Also, they have been grouped in their normal habitats. I have given the accepted English meaning of the Latin and Greek terms used in the textbooks as far as possible, and have expressed size in inches - the measurement familiar to weavers. The nomenclature is modern, but the names under which the lichens have been known in the past are given as well, together with the lovely and very descriptive names which the old dyers and country people gave to these plants.

The Orchil-producing lichens do not yield their dye by just boiling the plants - their pale-violets, strong purples and reds are obtained by the ammonia method. But there are many other lichens which give good browns and yellows to boiling water. Both kinds of lichens, as well as their dyes, are described. The methods for extracting Orchil are based on first-hand experiments. Several books on vegetable dyes refer to Orchil, but do not give any accounts of actual experience in making the red and purple dyes. Some even inform the reader that these colours were never obtained from British lichens, but only from imported plants! Textile designers could browse with great benefit among the infinite soft colourings and textures of the almost-unexplored territory of the vegetable kingdom.

Lichens have been described as 'humble little handmaids of Nature'! They are the most grasping, tenacious and dogged little plants in all the vegetable kingdom. They have a way of doing a vanishing trick in dry weather, so that one can walk about the countryside without realising their existence. When once they arrest the eye and hold one's interest, there is no more fascinating occupation than looking for them, while the high-altitude lichens will lead the searcher into some of the most glorious regions of the world.

I am greatly indebted to Mr Peter James of the Botany Department of the British Museum, for his lively interest and very sound advice. Also, to Mr A.C. Jermy for so kindly bringing the names in the manuscript up to date in the absence of Mr James abroad. Finally, in describing the distribution of lichens in North America, I must acknowledge the valuable information given by Mason E. Hale in his *Lichen Handbook* (Smithsonian Institute).

                                            Pont Eidda, North Wales
                                            1960

CHAPTER I

THE LICHEN DYES

The use of the lichen plants for dyeing seems to have originated in remote antiquity. The moors, highlands and sea-girt promontories of Britain, Ireland, Iceland and Scandinavia appear to have been inhabited by a people with a great knowledge of the plants, which were used both for medicinal purposes and for dyes. Traditional methods of obtaining colours, including reds and purples, have been handed down to our own day, together with the old and very descriptive names for the plants.

The Phoenicians in the Levant made their Tyrian purple from the Murex and Baccinium shell-fish - a practice which continued up to the twelfth century A.D. It was a royal and ecclesiastical privilege to wear clothes of this colour. Cloth receiving this dye was often first coloured with a preparation made from two lichens which abounded in the Levant, *Roccella tinctoria* and *Roccella fuciformis*. When the art of making the famous Tyrian purple died out, the lichen dye *alone* was used for making a regal purple. This preparation from lichens is referred to as Argol, Archil or Orchil.

The art of making Orchil cannot have been known in Italy before the fourteenth century, because a Florentine merchant seems to have come upon the method early in that century, while travelling in the Levant. On returning to Florence, he experimented in secret, and produced a dye for wool, selling it under the name of Tournesol. He later made his knowledge public, and as a result Italy long enjoyed the monopoly of making and selling Orchil.

An Italian dyer named Rossetti passed on the knowledge to England, and then the London Orchil-makers began importing, at great expense, the *Roccella* lichens from Africa, the Canaries and the Cape Verde islands. Later these were supplemented by other plants, including *Umbilicaria*, known as Pustulous Moss, from Scandinavia; *Parmelia perlata** from the Canaries; and quantities of indigenous British plants, including *Ochrolechia tartarea* and *O. parella*. These latter lichens were collected by country people in the north of England. In Scotland a similar preparation to Orchil was made in the form of a carmine powder called Cudbear, the lichens used being *Ochrolechia tartarea* and *Urceolaria calcarea**. From the same lichens the Dutch made another form (called Lac, or Litmus) which was sold as small cakes of a dark blue colour. In France the famous dye, Orseille d'Auvergne, was macerated from *Ochrolechia parella*. It was kept moist in casks, and was said to smell like violets.

One wonders who first discovered how to make the very unpromising-looking, white, mealy *Ochrolechia tartarea* run as rich and red as blackberry juice. Had the art of making these Orchil dyes come to our shores during the Bronze Age, was the knowledge brought by Crusaders, or did the art begin among the peasantry only after the method had been introduced from Italy? Traditional methods of making some form of Orchil have been handed down by word of mouth to our own day.

Originally, the only available form of the ammonia required for macerating Orchil was that to be found in putrid urine, so every crafter was able to make his own dyes; this commodity was collected from house to house for commercial use. It was also used for scouring wool.

Whole families must have been able to identify these lichens by sight, feel and traditional teaching. Today, anyone setting out to find the plants is at a loss to know where to begin. When once the lichens have been seen alive and growing, however, they are not easily confused with those that do not give a dye. It is hoped that the coloured plates in this book will in some way compensate for the loss of inherited knowledge in the gathering of these dye lichens.

Those lichens which give a dye to boiling water are still used in the Highlands, in Ireland and Scandinavia, and, until a few years ago, in Wales also. In Scotland crottle is the general term for dye lichens, and the plants are differentiated by being called either Black, Dark, White or Light Crottle. In Ireland the name Arcil or Arcel seems to apply to a number of dye lichens, which suggest that the older and more ancient name has survived. In Sweden, a lichen, known as *Candelaria* was scraped off the rocks and actually mixed in with the tallow for making golden festival candles.

Thomas Pennant, during his tour of North Wales in the eighteenth century, found that the inhabitants dyed their homespun cloth with the lichens *Parmelia omphalodes* and *Xanthoria parietina*. And recently, in the same part of the country, an old lady was still using both these plants. She seems to have steeped the *X. parietina* in urine, and then dyed with soda in the dye-bath, for she produced a kind of puce colour. It appears that a striped material, like taffeta, was made with the puce and the rich *P. omphalodes* brown dye. A red dye was also obtained by steeping *O. tartarea* in urine.

May and June have always been considered the best months for gathering the crustaceous *Ochrolechia tartarea*, while the two lichens *Parmelia omphalodes* and *P. saxatilis* should be gathered after rain in August. These latter are still used today in Scotland and the outer islands. They are either used fresh, or are carefully dried in the sun, and then stored in sacks.

When used fresh, *P. omphalodes* and *P. saxatilis* impart a lovely aroma to the wool. Many other lichens which, like these, only needed to be boiled, were used for both browns and yellows.

CHAPTER II

THE LICHEN PLANT

A lichen is composed of two plants, a fungus and an algae, living together in a very close, symbiotic relationship, and each supplying the other with certain necessary elements. This relationship can be seen by cutting a thin section through a foliaceous plant and examining it under a microscope. The fungal body of the plant will be seen as a white strip, with a dark outer surface above and below, while a little below the dark upper skin (or cortex) will be seen a thin band of green, this being a layer of green algal cells. In some species the algae are distributed among the fungal filaments or cells. The characteristic green hue of the lichen plant is due to this layer of algae shining through the upper surface when the plant is wet.

The fungal cells or filaments provide the fruits, which are generated in the body of the plant, which slowly emerge usually in the form of a minute pimple, on to the outer surface. The immature fruit gradually opens out and eventually takes on whatever form is characteristic of a particular plant. Except for *Peltigera canina* and *Umbilicaria pustulata*, most of the lichens given in this book have a rather button-shaped fruit.

Within the disc of the fruit, the spores in their case come to full maturity, and they are expelled through the tips of their cases or *asci*: the spores may then throw out filaments and lay hold on another algae in order to generate a new plant. Lichen fruits take a very long time to reach maturity, but they will continue to supply spores for many years.

There are two other ways in which a lichen plant can reproduce itself, both in the form of outgrowths on the surface. The presence or absence of these growths, as well as the way in which they are grouped, can often help in the identification of a lichen. The first of these is in the form of a granular powder, made up of minute bodies, each containing the fungal and algal cells. These are called *soredia*, and they can either germinate on the plant, or be scattered, and so form new plants. The other outgrowths are called *isidia*, and are made up of minute, coral-like projections. These differ from *soredia* by being covered with the upper cortex of the lichen, and can only germinate on the surface of the plant itself. Lichens have been divided into three main types or groups: (1) *Crustaceous*; (2) *Foliaceous*; (3) *Fruticose*.

(1) The *crustaceous* lichens form either a thick or thin crust on rocks, bark or soil. The lower cortex is usually absent, and fungal filaments are extended downwards into the substratum. This undergrowth can be either light or dark, and often gives a boundary between plants of the same or other species.

(2) The *foliaceous* lichens are leafy extensions of one or many lobes. The lower cortex is usually provided with hairs, or some form of holdfast, by which the plant adheres, either loosely or very closely to the substratum.

(3) The *fruticose* are *shrubby* lichens which continually branch from a basal sheath or holdfast. They can be erect or pendulous, flaccid or stiff, and there is great variation in their form and structure.

The crustaceous lichens are supposed to have been the first plants on this earth, corroding the rocks and so helping to create soil. They can withstand great variation in temperature, and are able to survive a drought by becoming inactive. The foliaceous lichens do not reach such high altitudes as the former plants, and in the hot and humid tropics they become large and exotic. In very arid countries these plants rely for their supply of moisture on the heavy falls of dew. Certain lichens prefer certain kinds of rocks or trees, but all need light and air, and will not survive near smoky towns.

There are a great number of chemical properties in the lichen plants, but the one we are concerned with is that which yields a dye. The lichen acids are the source of the colouring matter. As these acids are affected by the amount of light the plants receive, better results are likely from plants growing in sunny situations and from gatherings made after a good summer.

These lichen acids, of which there are several kinds, often give a plant its bright yellow or orange colour. *Xanthoria parietina*, for example, owes its rich yellow-orange hue to parietinic acid, and will give this colour to boiling water. This acid is affected by an alkali, and so one can get a puce or light plum colour by adding soda to the dye-bath.

A section cut through *Parmelia saxatilis* when wet will soon turn a red-brown colour, and the lichen acid seems to be in the form of minute brown specks on the outside of the fungal filaments.

The colourless acids are perhaps the most interesting - that called *orceine* giving the colouring principle of Orchil; and a plant will give no indication of possessing this acid unless tested with an alkali. The white, mealy *O. tartarea* shares the property of giving Orchil with the very dark, foliaceous *Parmelia fuliginosa*.

In order to find out what colour a lichen will yield, a little of the powdered plant should be shaken up in a small glass with a pungent solution of ammonia and water. The glass should be closely sealed for a day or two, after which the color will be apparent.

Gathering the Plants

Lichens are more easily gathered in calm and still conditions following rain. They almost vanish from sight in dry weather, and are easily broken and wasted when gathered in a brittle state.

If plants are being taken from a wall, it is best to work with the strongest light coming over one shoulder. Thin-bladed knives are most suitable for detaching those foliaceous plants which adhere very closely, but scrapers are best for the peat-like pads of *omphalodes* and for the rather mealy, crustaceous *tartarea*. Unless the plants are going to be used straight away, they should be very carefully dried in the sun and put into bags in a dry place. The crustaceous lichens should be dried and fermented straight away. There are many little mites in the cracks and crevices, and they go on spoiling the plants. In any case it is a good idea to make use of the summer warmth for fermenting these plants, which are normally gathered in May and June, while the foliaceous ones are not collected until the end of summer.

Most lichens take years to grow, so care should be taken to gather no more than is actually needed. Devon, Cornwall, the Lake District, the Pennines, Wales, Scotland and coastal regions in the Northwest, will prove the best hunting grounds for the handweaver in Britain.

Rare and Endangered Species

Much work has been done on the status of lichens since this book was first published. A lichen rare in North America may be common in the British Isles, and vice versa. Dyers are urged to seek assistance at herbaria and from specialists; to avoid over-collecting; to consult libraries for literature appropriate to their geographic location. [The Editors.]

Symbols

c common
e endangered
r rare; do **NOT** collect
B British Isles
NA North America

CHAPTER III

THE ORCHIL-PRODUCING LICHENS

Evernia prunastri - Ochrolechia parella - Ochrolechia tartarea - Parmelia fuliginosa* - Umbilicaria pustulata*

EVERNIA PRUNASTRI (L.) Ach. (*prunus*, a plum tree.)
The Stag's Horn or *Ragged Hoary Lichen*
See PLATE III, details on PLATE I.
In North America, most likely to be found in the temperate forest areas. Status: e/B; c/NA (west); e/NA (east).

Found on trees, twigs, wooden pales, bushes and in orchards. This lichen becomes very lively after rain, and is not unlike the antlers of a stag.

Plant in transition between a foliaceous and a shrubby or fruticose form. Grows from a basal sheath, in part erect or pendulous, flabby and soft when wet, consisting of unevenly strap-shaped fronds, continually branching and narrowing, the axils of fronds smooth, oval and wide. The upper side yellowish-green-grey, slightly convex, with depressions bounded by long taut ridges, which often have pale granules (*soredia*), underside slightly channelled, pure white, with upper surface in reverse; fruits rare, usually projected along upper side of margins, sometimes on face of plant, on short, stout pedicles, rims continuous with pedicles, thin, discs dark brown.

The fronds begin by growing out straight from the bark and then they become pendulous, fresh ones forming above and in turn drooping, until a tuft or cushion is formed. On some old plants the powdery granules (*soredia*) have germinated into numerous little fronds, and these older plants are usually more wrinkled and distorted.

Prunastri will encircle the black twigs of the wild plum with lacelike ruffs or collars, and the pink buds opening above can look delightful. It will clothe the weather side of trees with a shaggy coating. This plant can be distinguished fom the rather similar *Ramalinas* by the pure white undeside, and the smooth oval axils, where the fronds branch out. As the *Ramalinas* often grow with *E. prunastri*, the two plants can be confusing.

A handful of freshly gathered moist plants of *E. prunastri* have a strong smell, similar to chloride of lime. It was used at one time for the gum it yields, and in Egypt for flavoring bread.

E. prunastri gives a deep plum colour with ammonia on wool.

OCHROLECHIA PARELLA (L.) Massal. Previously known as *Lecanora parella*.

The Crawfish or *Crab's Eye lichen. Light Crottle.*
See PLATE II.

In North America *Ochrolechia pallescens** may be more common. It is similar, and also yields a red dye. Status: c/B; r/NA.

Found on stones in old walls and on rocks.

Plant forming a wide-spreading crust, generally of a definitive outline where it is thicker and white, the undergrowth white; crust thickish or thin, smooth or deeply cracked, the areas separate; colour whitish or greenish-grey, greener when wet; usually very fertile, the fruits often pressed close together; the discs at first concave, becoming plane, then convex, pale flesh-coloured or whitish, and covered in a white bloom; rims thick, smooth and prominent; young fruits almost clothed with swollen rims, fruit moderate in size.

This lichen is called *Lecanora parella* in books on vegetable dyes. The appearance of the fruits may have given this plant the name of the Crawfish or Crab's eye lichen: the young fruits with their almost closed-up discs and swollen rims do suggest crab's eyes, especially when they are leaning against each other.

This was the lichen gathered in France under the name of Perelle d'Auvergne, and it was used for making the dye Orseille d'Auvergne. It was used in the Highlands to obtain a reddish or orange dye, and quantities were gathered in the north of England for the London Orchil-makers. *O. parella* is a much thinner crustaceous plant than *tartarea*, and usually very fertile, with smaller and rather prominent fruit. It loves old stone walls, river bridge parapets and boulders near streams. There is a white and a more grey variety, the latter being considered the best for making the dye.

The writer has obtained a soft claret red by macerating with ammonia.

OCHROLECHIA TARTAREA (L.) Massal. Previously known as *Lecanora tartarea*.

The Cudbear Lichen. Crotal, Crottle, Corcir, Korkir (Scotland), *Ormassa* (Norway and Sweden).
See PLATE IV.

In North America, *Ochrolechia tartarea, var. frigida*, may be more common, and will also yield a red dye. Status: c/B; r/NA.

Found on rocks, boulders and old trees in upland districts.

Plant forming a crust, wide-spreading, thick, friable, mealy and tartareous, very uneven, of crowded granules or warts, whitish-grey or white, greener when wet; raised parts of younger plants tinged yellow-green, older crusts with deep cracks in them; plants mostly sterile: when fertile, fruits very numerous, crowded and often large, a little under ½ inch across, the discs pinkish-buff colour, a little corrugate, the rims thin and inflexed; several fruits may join up and form a four-petalled disc.

This is usually known as the Cudbear lichen in books on vegetable dyes.

O. tartarea is one of the thickest and most uneven of the crustaceous lichens. It may grow in roundish patches, or be spread out along the ridges and laminations of shaly rocks. It is very friable and easily detached from the rock, and sometimes it grows over moss. An eighteenth-century encyclopedia describes it as looking like meal clotted together, and young plants have the texture of close-grained cauliflower heads.

The country people of Northern England and Scotland steeped the powdered *O. tartarea* in putrid urine for many weeks then, when it had reached the necessary purple or red stage, they made the mass up into balls with lime, chalk or burnt shells, and hung them up to dry. When needed, this form of Orchil was powdered and boiled up with a little alum. It was used in Wales in the same way for a red dye.

This lichen must be at least five years old before it is gathered.

PARMELIA FULIGINOSA* (L.) Nyl. (*fuligo*, soot)
See PLATE IV and Details on PLATE I.

In North America, found on rocks, and on shrubs and trees; in the British Isles, on smooth stones in walls and on rocks under foliage. Status: e/B; c/NA.

Plant foliaceous, very thin and tenuous, adhering very closely, usually about 1 ½ inch in diameter, dark brown, with dull matt surface in centre of plant, glossy and smooth at circumference; plant much greener when wet, especially the circumference; centre of plant (*lens*), lengthening towards centre of underside; fruits infrequent, small, scattered, discs brown, rims thick and wrinkled and bearing one or two outgrowths.

P. fulginosa is not often mentioned as a dye lichen, but the writer has experimented with it, and it yields a good Orchil. It does not grow in sufficient quantities for general use in the British Isles, but for small skeins of wool or silk it is worth trying, and the colour seems to be fast to sunlight.

This lichen forms dark brown spots on walls, but many plants may join up and cover a large stone. The plant seems to like a damp situation under a hedge or along a woodside, and it prefers smooth, clean stones. Another variety, *P. laeteivirens**, which is not dulled in the centre by outgrowths, may be found on trees in woods. Both these lichens will given an Orchil by macerating with ammonia.

Fulginosa is so closely attached to the stone that it needs a very thin blade to remove it. It is so tenuous and thin, that it is impossible to gather it in even the slightest breeze.

UMBILICARIA PUSTULATA (L.) Hoffm. (*pustulata*, a pock or blister).
Pustulous moss. Rock Tripe.
See PLATE V and details on PLATE I.

In North America known as *Umbilicaria papulosa*, or *Lasallia papulosa* (Ach) Llano. All the Umbilicarias, once known as Gyrophoras in Britain, will give Orchil, except *var. cylindrica* which, according to Lindsey, gives a greenish-brown colour. Found on rocks in Great Smoky Mountains, Southern Rockies, Appalachians, Great Lakes and along the East Coast, on granite rock. Status: r/B; c/NA.

Found on rocks in mountainous districts and coastal regions.

Plant foliose, of one leaf, thickish, held below by one stout, short root or holdfast, becoming large (up to 4½ inches) in diameter, and irregularly torn; when wet, it is bright olive green, shiny, flabby and a little gelatinous; when dry, it is grey-brown, tough and leathery, with a dense bloom on upper side; upper surface with large oval blisters or pustules, set among smaller and crowded ones, all radiating from the navel or centre of plant, which is a little depressed and also smoother; upper surface often bearing clusters of tall greeny-black outgrowths (*isidia*), silky when wet, stiff and harsh when dry; this side also (*lens*), with a mesh of very fine (in parts, coarse) cracks, the separate areas of which have a bloom; underside of plant with oval cavities corresponding to the pustules on upper side, like a honey-comb, the cavities bounded by smooth taut surface radiating from the central holdfast; this surface (*lens*) also with coarse (sometimes finer) cracks, the separate areas with a faint bloom in parts; fruits very rare, discs plane, elongate or oval, brown, with a black, rough margin, without a rim.

Fruits are extremely rare in Britain, but the writer has found one or two very small ones, and a description has been given together with a drawing of a spore on the Detail Plate. When gathered wet, *U. pustulata* feels and smells rather like seaweed. In order to conserve these plants they should never be gathered when fertile, nor should more be taken than are really needed. These Orchil-producing lichens go much further than those which are merely boiled. This particular lichen must be very carefully dried, for it attracts moisture like seaweed, and was used at one time as a hygrometer. A black paint used to be extracted from this plant, and it has been eaten in dire necessity by Arctic explorers, to whom it is known as 'Rock Tripe.'

U. pustulata yields a good red by macerating with ammonia. It has more brown in it than the red from *O. tartarea*, and a little more body.

CHAPTER IV

THE BOILING-WATER LICHENS

Hypogymnia physodes* - Lobaria pulmonaria - Parmelia caperata* - Parmelia conspersa* - Parmelia saxatilis - Parmelia omphalodes - Peltigera canina - Platysma glaucum* - Xanthoria parietina.

HYPOGYMNIA PHYSODES (L.) Nyl. Previously known as *Parmelia physodes*.

Dark Crottle. *Bjork-laf* (Sweden).
See PLATE III and details on PLATE I.
In North America, found in Appalachians, Great Lakes, Northwest, Pacific, Eastern States. Status: e/B; c/NA.
Found on trees, twigs, palings and to a lesser degree, on rocks.

Plant foliaceous, wide-spreading, very variable in size and form, the whole plant hollow and swollen looking, adhering in places by a very tenuous lower skin or cortex; when wet, glaucous-grey and very elastic, the lower cortex easily rubbed off; when dry, silver-grey and stiffish; lobes narrow, sinuous, continually dividing, extremities often blunt; younger lobes inflated at the tips; older lobes with the lower cortex ruptured, the tips springing upwards, revealing tissue of plant, and on these white lipshapes clusters of a powdery outgrowth (*soredia*), upper surface smooth with colonies of minute black specks towards lobe ends; underside of plant very black with a sheen, except at circumference which is a light brown, often with holes beneath older lobe ends, wrinkled and puckered in centre at points of adhesion; fruits developing as pimples at lobe-ends, rising up on pedicles, which are also hollow, discs moderate in size, becoming large with age, rims thin and inflexed, discs greenish-brown or light brown.

This is a very curious lichen, for unlike most of the foliaceous plants, the lower cortex is separate, and the lobes hollow. As *H. physodes* clings by this very tenuous underside, it is always falling off branches in wild, wet weather, and pieces lying about among dead leaves can look very green. Small plants on larch twigs seem to lose the lower cortex altogether, except where they adhere.

H. physodes is distinguished by the lip-shaped lobe-ends which become covered with clusters of powdery granules (*soredia*). Other varieties will also give a golden brown to boiling water.

Plants on rocks will sometimes be so smooth and continuous as to appear almost of one leaf, the pattern of the lobes being merely embossed on the surface. On meeting with any unevenness in the substratum, however, the older lobes tend to rise up and separate, developing the typical re-curved lobe ends.

Plants found on kindling wood which has fallen from a good height are often very fertile, mature fruits being large with deep upstanding rims.

LOBARIA PULMONARIA (L.) Hoffm. (*Pulmo*, the lungs). Previously known as *Sticta pulmonaria*.

Rags, Lungwort, Aikraw, Hazel-raw, Oak-rag, Hazel, Crottle.

See PLATE III and details on PLATE I.

In North America, the temperate and boreal regions, as of most continents, share this lichen. Status: e/B; c/NA.

Found on old forest trees, especially oak and maple; rarely on rocks.

Plant foliaceous, large, thickish, wide-spreading and branching, olive green and flabby when wet; yellowish brown and stiff when dry; upper surface with large network of branching, taut, veinlike ridges, enclosing deep depressions, the veins often strung along with pale, bead-like granules (*soredia*), veining finer and closer and more yellow at lobe-ends; lobes broad, elongated, branching, sinuous, with wide, rounded indentations, the axils of the branches often forming wide, circular openings through plants; lobes ending abruptly, square-cut, with rounded, shallow notches; underside of plant deep ochre colour and downy, with hump-like swellings corresponding to depressions on upper side, lighter and less downy over the humps; when fertile, fruits usually on margins of upper side, at first closed, then open, discs brown with lighter rims.

This lichen is called *Sticta pulmonaria* in books on vegetable dyes.

L. pulmonaria is easily seen when dry, being then of a yellowish-brown colour, which shows up against a mossy tree trunk. The plant seems to favour the oak, although it can be found on maple and beech, near streams and rivers.

Old, rather ragged plants, with much granular out-growths (*soredia*), can look very like torn black rags that have gone green with age.

At one time *L. pulmonaria* was prepared as a jelly, and given to those suffering from pulmonary affections.

The Herefordshire country people used to dye their woollen stockings brown with this lichen. It was also used in the Lowlands of Scotland, as one of the crottles, and in Northern Ireland - where it is called Hazel-rag.

L. pulmonaria will dye a good auburn and rusty brown colour on wool with boiling water.

PARMELIA CAPERATA* (L.) Ach. (*Capero*, to wrinkle).
Stone Crottle, or *Acel* (Ireland).
See PLATE III and details on PLATE I.

Found in temperate forest areas, mostly on Red Oaks, to a lesser degree on White Oaks, and on sandstones. In North America, common on old trees. Status: e/B; c/NA.

Found on trunks of old trees, palings and boulders.

Plant foliaceous, thickish, large, wide-spreading on oval patches, pale yellowish-green, surface often powdery and granular in centre of plant (*soredia*), and more dingy in colour; lobes broad (up to ½ inch), rounded, rather concave with flatter margins, edges scalloped and narrowly notched; young lobes smooth and shiny, older ones finely wrinkled and puckered, either across or lengthwise, away from the extremities of the lobe-ends; minute lobules in central, granular area, very conspicuous by contrast; underside black but becoming light brown and shiny at circumference; hairs begin some way in from lobe-ends as white pimples, then sprouting as white hairs (*lens*), and becoming black towards centre of underside; fruits infrequent, central, moderate in size, discs brownish-red, rims notched and granulose.

Even from a considerable distance, this lichen will shine out on the bark of a tree. It usually grows in large, rather oval patches loosely attached; stiff when dry, but barely loses its colour. This may be because the fungal cells (which may be the source of the yellow dye) immediately below the upper surface are yellow.

The very fine wrinkling and puckering on the older lobes distinguish this *Parmelia* from others, as do the white hairs on the underside which may be seen under a lens or microscope. Caperata grows on a variety of trees and on rocks. In Ireland it is called Stone Crottle or Arcel, being used for a yellow dye on wool. It is said to yield orange and brown also. With boiling water it gives a good clear yellow.

PARMELIA CONSPERSA* (Ehrh.) Ach. (*Conspergo*, to besprinkle).
One of the Crottles.
See PLATE V and details on PLATE I.

In North America this *Parmelia* is found on sandstones, on quartzite and granite in open woods, pastures and fields. Status: c/B, c/NA.

Found on stones and boulders in highland regions in Scotland, coastal regions in Ireland.

Plant foliaceous, orbicular, thin, very closely adhering, sometimes wide-spreading and irregular; centre of plant grey-green and polished-looking, circumference light yellow-green; outer lobes widening towards their ends, small, rounded, sinuous and notched; lobes smaller and more branched and crowded towards centre of plant; underside dark brown with short hairs almost to lobe-ends; fruits sprinkled all over central area, young ones towards the circumference, older fruits in centre of plant, between them (*lens*) minute rod-shaped outgrowths (*isidia*), some already germinated into minute lobules; young fruits small, thick-rimmed and close-pressed, mature fruits often large, very slightly top-shaped, held below at centre, discs chestnut brown and corrugate, rims thin, smooth and inflexed, fruits often run together.

P. conspersa seems to like a warm sunny rock-face at an altitude of 800 feet and upwards; and the whole plant adheres very closely, needing a thin blade to remove it.

After a heavy shower of rain, when the sun has come out again, the brown fruit discs turn a vivid viridian green. The grey-green centre of the plant darkens too, and merges into the stone, so that the rock-face appears to have only rings and circles of yellow-green left growing upon the surface.

This lichen (formerly considered one of the Crottles) gives a good brown to boiling water on wool.

PARMELIA SAXATILIS (L.) Ach. (*Saxum*, a stone or rock).
Crottle, Staney-raw (Scotland), *Scrottyie* (Shetland), *Sten-laf, Stenmossa* (Norway and Sweden).

See PLATE II and V and details on PLATE I.

In North America this *Parmelia* is found on sandstones, quartzites and granites. On acidic gneisses in New England. Status: e/B; c/NA.

Locally common in highlands and on mountain talus, occasionally on trees but increasingly vulnerable to depletion by dyers.

Plant foliaceous, orbicular or wide-spreading, thin, adhering closely except at circumference, greyish-white or glaucous green, greener when wet; upper side covered with a network of mesh of very fine ridge and depressions (*reticulations*), closer and finer towards the lobe-ends, the ridges near surface white, often cracked open and appearing dark, sometimes with outgrowths (*isidia*); lobes narrow, clean-cut, sinuous, notched and sharply incised, lobe-ends square-cut, and incised, also browner and upturned; minute lobules may have germinated on pulverent central area; underside very black with dense black hairs right out to the lobe-ends; here and there (*lens*) small white circular pits or depressions, showing tissue of plant; fruits infrequent, merging from granular area on the upper surface, moderate in size, discs reddish-brown, rims rather thin, often granular and notched.

This lichen merges into rocks or stones when dry, but is greener and more obvious when wet. Some plants are tinged a light bronze colour.

A young plant on stone is usually orbicular, and the sinuous lobes radiate with a great feeling of movement, almost as though they were turning a little clockwise as they grew.

P. saxatilis will often spread and cover large boulders by the joining up of several plants. Often the centres of old plants will peel and slough off a stone, leaving only the circumference still growing. Fruits are reddish-brown but not commonly found.

P. saxatilis is one of the Crottles still in use for dyeing wool a strong reddish-brown or dead bracken colour. It is a very fast dye, and leaves a pleasant aroma in the wool.

Alder bark and *Xanthoria parietina* used to be added to the dye-bath as well.

PARMELIA OMPHALODES (L.) Ach.

Cen du-y-cerrig (Wales), *Black Crottle, Cork Corker, Crostil* (Highlands, *Arcel* (Ireland), *Alaforel leaf* (Sweden).

See PLATE II and IV and details on PLATE I.

In North America found in the same areas as *Parmelia saxatilis*, but at higher altitudes. Status: e/B; r/NA.

The lichen seems to grow best from about 800 feet and upwards, where it will spread and cover large boulders with peaty pads, or cushions of dark brown. Towards the centre of older plants, layer after layer will have been built up, the new surface lobes being very small and closely packed.

Plant foliaceous, usually wide-spreading, consisting of numerous narrow, smooth, shining and dark brown or purplish black lobes, crowded and overlapping; at the circumference of plant the lobes adhere more closely and are greener; whole plant forms flattish pads or cushions, usually spreading over boulders; plant much lighter and greener when wet; lobes narrow, clean-cut, incised, square-cut at extremities, with incisions; upper surface with a network of fine ridges and depressions (*reticulate*) much less marked than on *P. saxatilis*, and more diffuse, the white ridges cracked open, giving speckled appearance to lobes; never has outgrowths, either *isidia* or *soredia*; underside very black with dense black hairs right out to extremities of the lobes; fruits infrequent, when present numerous, up to 3/4 inch across, many may join up; discs red-brown, rims thin, lightish and speckled and inflexed. Found on rocks and boulders in coastal, upland and mountainous areas.

This plant was once considered to be a variety of *P. saxatilis*, which it resembles in the way that the lobes are incised and square-cut, and to a lesser degree, in the surface texture (*reticulations*). The fruits on *omphalodes* are much larger and more red-brown.

P. omphalodes gives a rich red brown to boiling water, and is one of the fastest dyes known. It imparts a lovely and permanent aroma to the wool when used fresh and, together with *P. saxatilis*, was one of the chief Crottles of the Highlands. Both these lichens are said to yield a red and purple dye.

This lichen is uncommon in North America, except in the far north. In the past it was often over-collected by dyers in those areas where it occurs in Scotland and Ireland.

PELTIGERA CANINA (L.) Willd. (*Canis*, a dog).
Ash-coloured Ground Liverwort.
See PLATE II and details on PLATE I.
In North America found in all temperate forest areas.
Found over turfy rocks and tree bases or stumps, and on moss.

Plant foliaceous, radiate in expansion, wide-spreading, of large lobes turning this way and that, somewhat overlapping and downy on upper side; when wet, dark greenish-brown, cool and crisp to the touch; when dry, a pallid ashen colour, sometimes tinged brown, shrivelled and papery; lobes long (up to 3 3/4 inches), widening to a broad, rounded apex; upper surface downy, undulating, the margins curved backwards, wrinkled, wavy and notched; underside without a cortical layer, white and glistening when wet, with roundish pinkish veins continually branching upwards, closer and finer at margins; from these veins spring long white fibrils, often visible from upper side; depressions between veining, corresponding to undulations on upper surface of plant; fruits adnate on upper side of margins, resting immediately on tissue of plant; discs bright chestnut colour and shiny, orbicular but becoming elongated by revlution of the margins, without rims.

This lichen has large lobes, mature ones being roughly the size of an oak leaf. This plant feels so alive when moist, and it clings so tenaciously by the long white fibrils, that one hesitates to remove it. It is very cool to handle, and the change to the dry state is always rather startling. First of all the veining begins to turn an ashen hue, and it is at this stage that the plants look rather like tabby kittens clinging by their white claws. Later the whole plant becomes ashen grey, withered and papery.

The immature fruits can be felt as pimples on the extreme edge of the turned-back lobe-margins. As they develop, they feel like tiny hooks, due to the disc being rolled up. As the disc flattens out it ruptures a thin veiling, and the bright chestnut colour can then be seen: at the same time the margin of the lobe comes over to the front. At all final stages of development the fruits resemble finger- or toe-nails in form.

P. canina will fringe great boulders where they emerge from turfy ground, along with smaller and more fertile varieties that are not downy. It will cling to grass and mosses in road banks, and grow over tree-stumps in damp woods. This lichen was used at one time as a supposed cure for canine madness.

Linen that has been mordanted with alum can be dyed yellow with this lichen.

The use of vegetable dyes is popular once again in the United States, and an American handbook describes a *Peltigera* (unspecified) used for obtaining either a yellow-tan with alum, or a rose-tan by passing wool through a hot solution of chrome and acetic acid. There is also *Peltigera polydactyl*, with a smooth, shiny surface, very fertile, dark green when wet; light, cold grey when dry and not downy), which might be promising to handweavers. It yields a soft pink with ammonia.

PLATYSMA GLAUCUM* (L.) Nyl. (*glaucus*, bluish-grey). Previously known as *Cetraria glauca*.
See PLATE IV.

Cetraria juniperina, which gives a fine yellow, is to be found in Southern Oak and Pine forests. *Cetraria glauca* will be found in the northern States, north Wisconsin, Michigan, New York, New England. Status: c/B, c/NA.

Found on tree branches and twigs, wooden pales, over moss and on stone walls.

Plant folicaceous, adhering in clusters by the underside, often widespreading, thin and delicate to the touch when wet, and a little yellowish; when dry, blue-grey and withered, the lobes rolling up; outer lobes very broad, the margins incised, crisp and upstanding; the upper surface sometimes with wide shallow depressions and ridges (*reticulate*), inner lobes shorter, more divided, narrowly waved at margins and upstanding; ridges and margins of lobes often with powdery outgrowth (*soredia*), which easily rubs off when wet; underside white or light brown, black and puckered at centre of plant; fruits rare, when present on extreme margins of central lobes, discs dark reddish brown.

This lichen has been included as it will give a yellow to the wool with boiling water. Another variety, usually known to dyers as *Cetraria juniperina*, is the lichen used in Sweden to obtain a beautiful yellow dye. It is a smaller plant than *glaucum*, and of a gamboge yellow on both sides.

P. glaucum can often look yellower than the name might suggest, especially when wet. It is a delicate little plant, and growing over moss can look rather like lettuce seedlings. A variety growing on the Alder tree is tinged brown, almost as if it had absorbed some dye from the bark. It is sometimes so coated in a powdery outgrowth that it loses all its fresh, crisp quality.

Some of these yellow lichen dyes might be used to bottom an Indigo dye, as they do not need a mordant, and can give very soft greens.

XANTHORIA PARIETINA (L.) Th. Fr. (*paries*, a wall). Previously known as *Parmelia parietina*.

Common Yellow Wall Lichen. Wag-massla. Wag-laf, (Sweden).

This plant has not been illustrated in the coloured plates, because the very distinctive golden colour of this lichen makes it fairly easy to identify.

In North America this plant is common in cemeteries, and near the sea, where bird perching rocks may be thickly covered in a loose, orange crust. Status: c/B, NA.

Found in cemeteries, on walls by the sea, and on trees and roofs in farm-yards.

Plant foliaceous, orbicular, wide-spreading, of a deep orange-yellow colour, green or blue-green in shade, adhering very closely; lobes very varied, small, rounded, undulating and rather concave at circumference of plant, towards centre, lobes as indeterminate lobules; upper surface smooth when fertile, or granular with *soredia* when sterile; underside white or pale yellow, smooth with occasional hairs or fibrils; when fertile, fruits scattered, small, discs deeper in colour, rims upstanding, wrinkled, and colour of upper surface.

X. parietina appears to like a salty atmosphere near the sea, and also the vapours from farm muckyards. Farm buildings and trees will often be covered with this beautiful plant, while a cottage quite near will have none.

This lichen seems to hold the sunshine even on the dullest of days. The acid which it contains is so affected by light that it is best to gather the plant from the sunniest places. In the shade, *X. parietina* will be a bluish-green or greenish-yellow.

The circular patches may spread widely, but often the centres will go, leaving the circumference still enlarging on the wall.

X. parietina seems to be full of possibilities; it will give a golden-brown to boiling water; on wool that has been macerated with ammonia, it may be used to obtain more of a puce shade, provided a little soda is added to the dye bath.

The writer has succeeded in getting a very lovely blue - a colour previously suggested in connection with these plants - from this lichen. This experiment with *X. parietina* is described in the chapter on Boiling Water Methods. The term 'blue' has always been used in the past to cover a wide range of bluish-purples from Orchil dyes. One is often told that the maceration can be stopped at 'blue' or continued to red.

LESS PLENTIFUL LICHENS WHICH GIVE DYES

Many of these species are rare or endangered now in North America and the British Isles.

Lecanora atra, the black-nobbed dyers' lichen. This is a grey or whitish-grey crustaceous plant, of crowded warts with a thin black undergrowth. Fruits are small and numerous, with jet-black discs and prominent rims. It makes Orchil.

Lecanora calcarea,* formerly known to the Cudbear-makers as *Urceolaria calcarea.* Said to yield scarlet, and was often mixed with *tartarea.* It is a white or greyish crustaceous plant, chalky or tartareous, with a white undergrowth, the fruits being very small and immersed, and it grows on limestone rocks.

All the *Diploschistes* species in Britain will yield Orchil. They are crustaceous plants. *D. scruposa** will yield a deep red dye. A number of white or light grey crustaceous rock lichens will yield dyes, especially *Pertusaria.*

Ramalina scopulorum. A rigid, narrowly divided, grey-green shrubby plant, growing on rocks and walls by the sea, that will give a yellow-brown to boiling water.

Haematomma ventosum. The red-spangled tartareous lichen. It is a finely-cracked, crustaceous plant, divided into cushiony areas, pale olive or greyish-yellow in colour. The numerous red fruits are very small, with light rims. Mr. Lightfoot, the botanist, had suggested that it might yield the same sort of colouring matter as *L. tartarea* or *L. atra.* Do not mix with other dye lichens, for the reaction with ammonia is entirely different. It first of all strikes a vivid green colour, which rapidly changes to a strong yellow, and finally to brown. It dyes a chocolate brown on wool.

Most of the *Usneas* will give good clear yellows to boiling water. *Usnea barbata** was steeped in stale urine for three months in Pennsylvania to yield orange dye. These plants are erect or pendulous, of rounded filaments, and yellowish-green in colour. They form a shaggy coating on old trees

Cetraria islandica. This plant was used in Iceland for a brown dye. It is a pale to dark brown strap-shaped shrubby plant, with spinulose margins, growing on the ground in most highland and Arctic regions.

Color Plates

The following five color plates are reproductions of Eileen Bolton's original drawings used to illustrate her first edition. She made these drawings of the lichens in their natural surroundings. The plants have been grouped so that their relative sizes, colours, and textures can be compared. The Welsh top-snails, bracken-shoots, butterflies, etc, have been included to give scale to the plants, for they are mostly rather small. Many distinguishing features, too small for the eye to see, have been enlarged on a separate Details Plate, together with any spores which have been observed.

Plate VI is a copy of the Frontispiece photograph which appeared in the first edition. Though this photograph reproduction is lacking clarity, it is an accurate representation of the colors acheived from lichen dyeing.

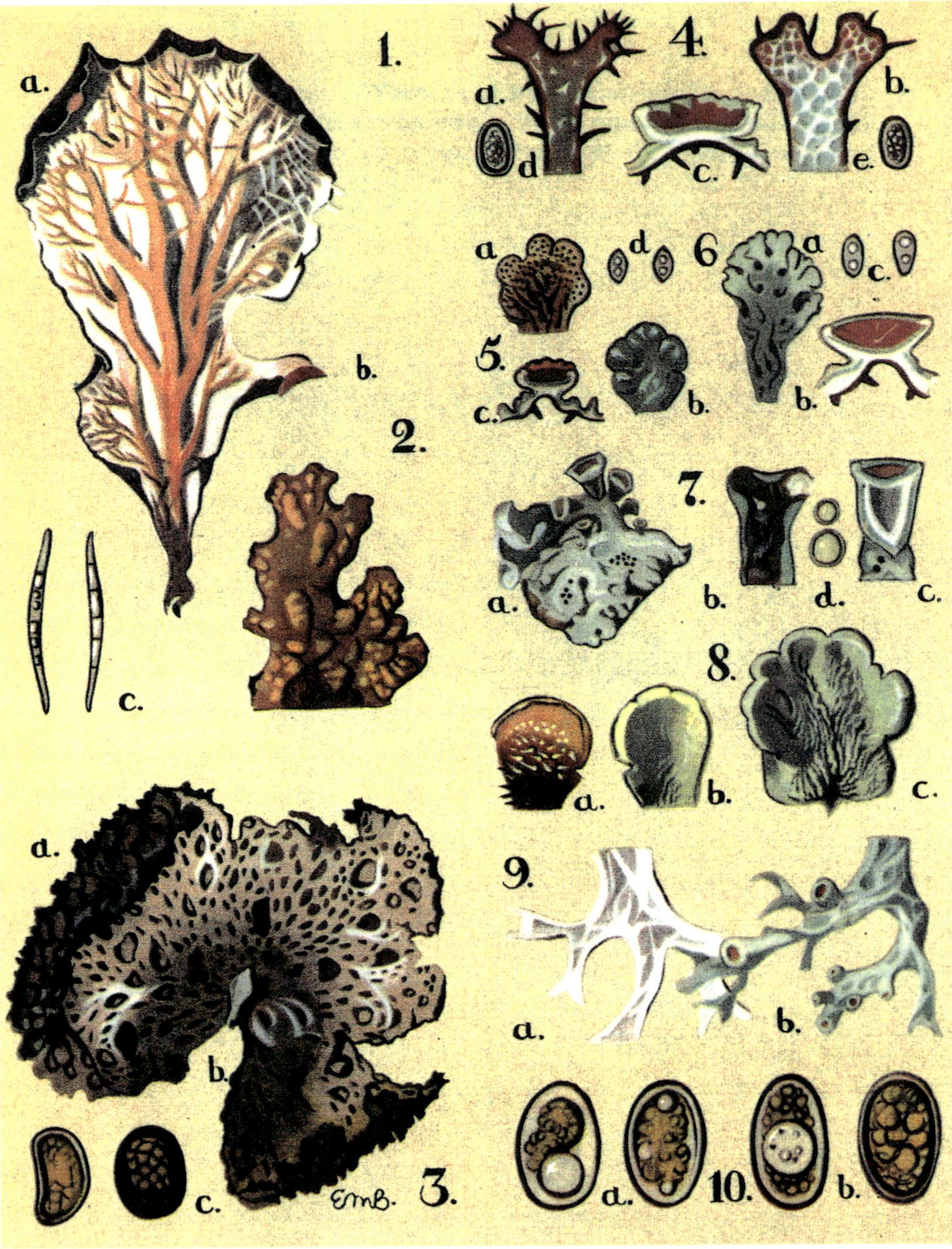
1.
a.
b.
c.
2.
d.
b.
c.
EMB.
3.
4.
a.
b.
c.
d
e.
a
d
6
a
c.
5.
c.
b.
b.
7.
a.
b.
d.
c.
8.
a.
b.
c.
9.
a.
b.
a.
10.
b.

PLATE I

Details of the Lichen Plants

1 *Peltigera canina*, underside. *a.* immature fruit on margin of lobe. *b.* underside of mature fruit. *c.* spores.

2 *Lobaria pulmonaria*, underside of plant.

3 *Umbilicaria pustulata**, underside. *a.* outgrowths on upper side, which is turned back. *b.* central holdfast. *c.* spores (microscope).

4 *Parmelia omphalodes* and *P. saxatilis* compared. *a.* lobe end of *omphalodes*. *b.* lobe end of *saxatilis* (much enlarged). *c.* combined fruit. *d.* and *e.* spores.

5 *Parmelia fuliginosa**. *a.* underside of lobe (enlarged). *b.* fruit in section (microscope). *c.* upper side of lobe. *d.* spores (microscope).

6 *Parmelia conspersa**. *a.* lobe upper side (enlarged). *b.* fruit in section. *c.* spores (microscope).

7 *Hypogymnia physodes*. *a.* complex lobe showing young and older sorediate lobe ends, also black specks or *pycnidia*. *b.* underside of lobe showing the lower skin ruptured. *c.* fruit in section (all enlarged). *d.* spores.

8 *Parmelia caperata**. *a.* young lobe on underside, showing the development of white hairs (*rhizines*). *b.* upper side of lobe. *c.* older lobe with wrinkling of surface (all enlarged).

9 *Evernia prunastri*. *a.* underside. *b.* upperside with fruits.

10 *Ochrolechia tartarea* and *O. parella*. *a.* spores of *O. tartarea*. *b.* spores of *O. parella*.

*Asterisks following names indicate lichens that have been reclassified (renamed); consult pages 39-41.

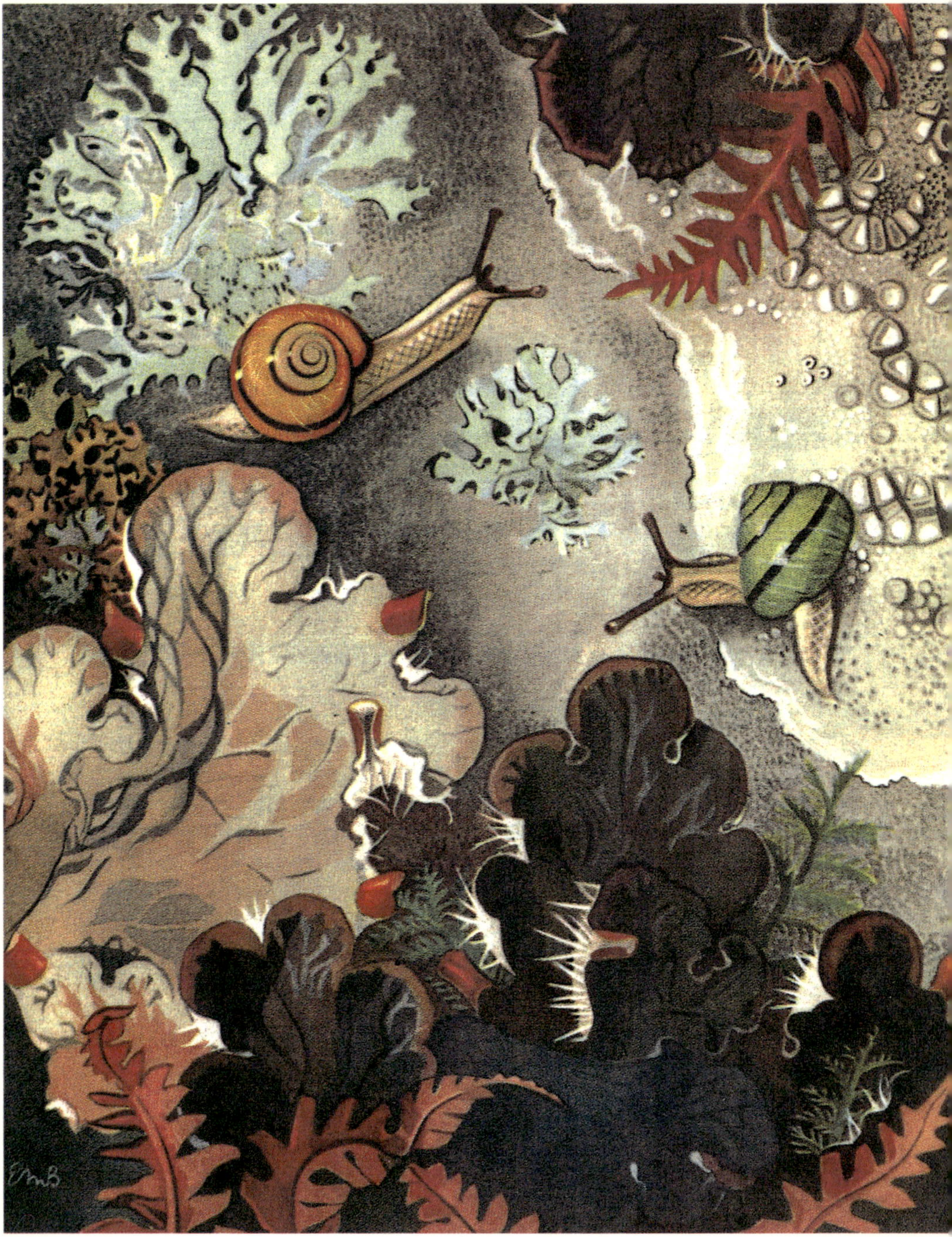

PLATE II

Parmelia saxatilis
(young)

Ochrolechia parella

Parmelia omphalodes
(young)

Parmelia saxatilis

Peltigera canina
(dry)

Peltigera canina
(wet)

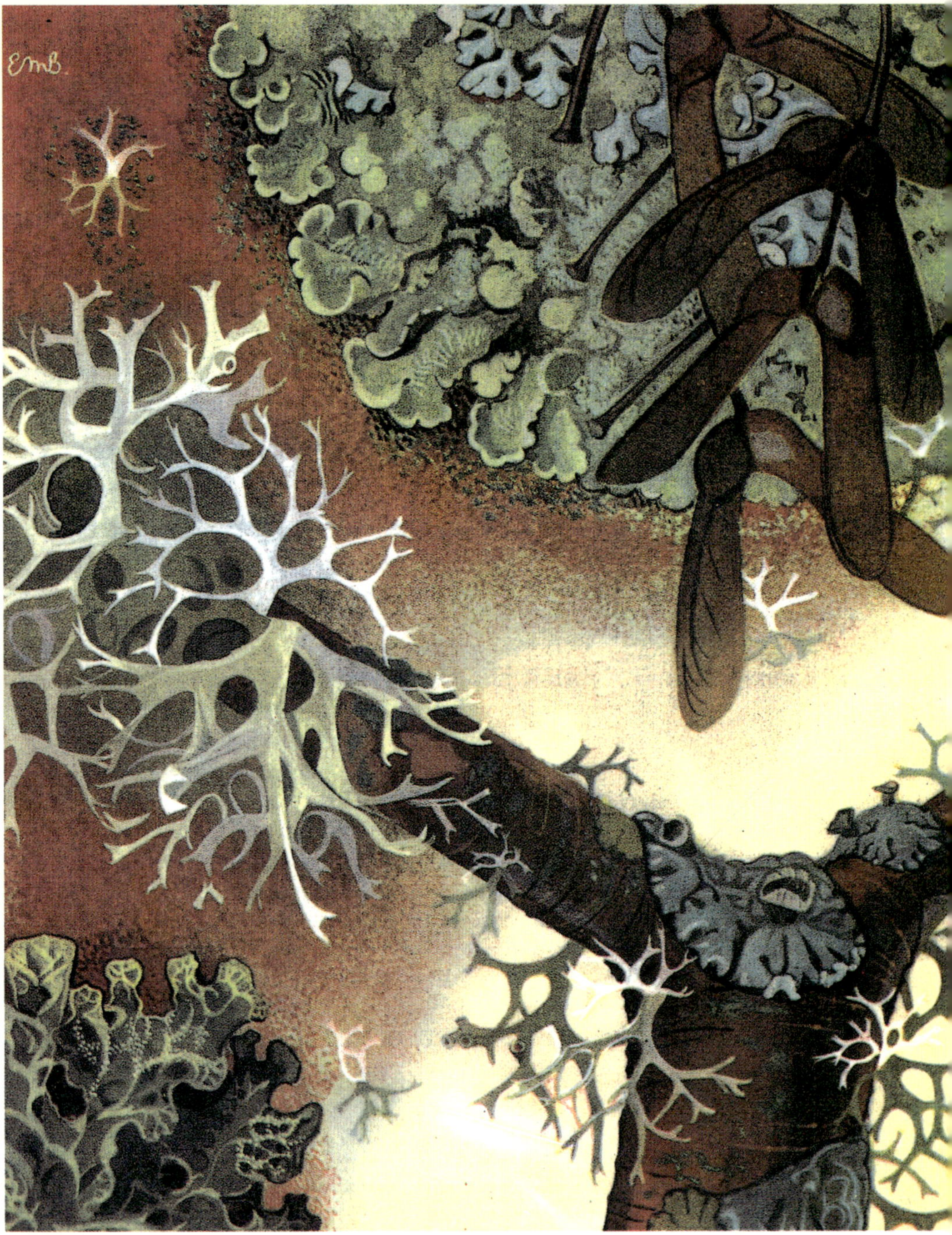
EmB.

PLATE III

Hypogymnia physodes
(small pieces, wet)

*Parmelia caperata**
(wet)

Evernia prunastri
(wet)

Hypogymnia physodes
(wet)

Lobaria pulmonaria
(wet)

Evernia prunastri, in fruit

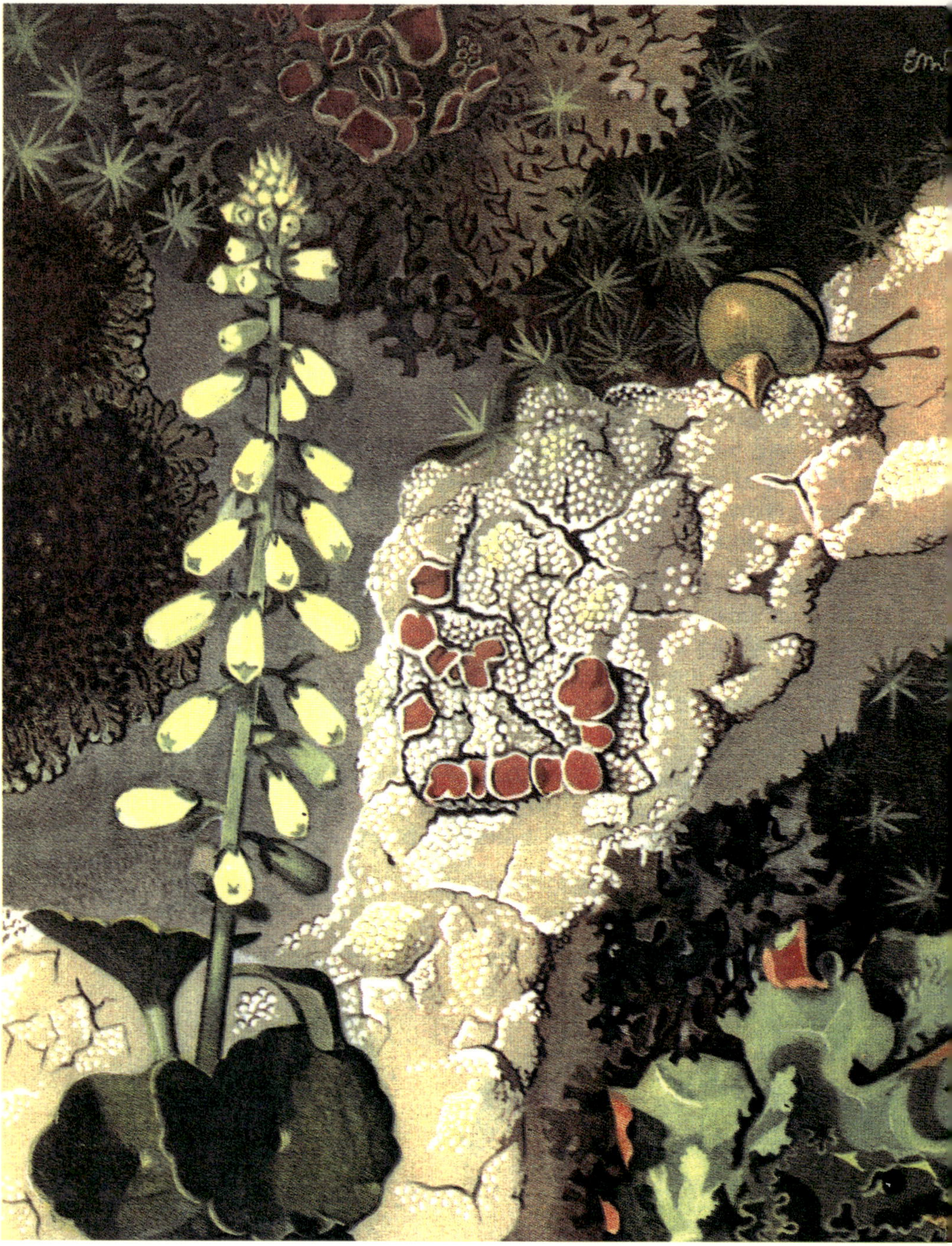

PLATE IV

Parmelia omphalodes
(wet)

*Parmelia fuliginosa**
(wet)

Ochrolechia tartarea
(old plant)

Parmelia omphalodes

*Platysma glaucum**
(wet)

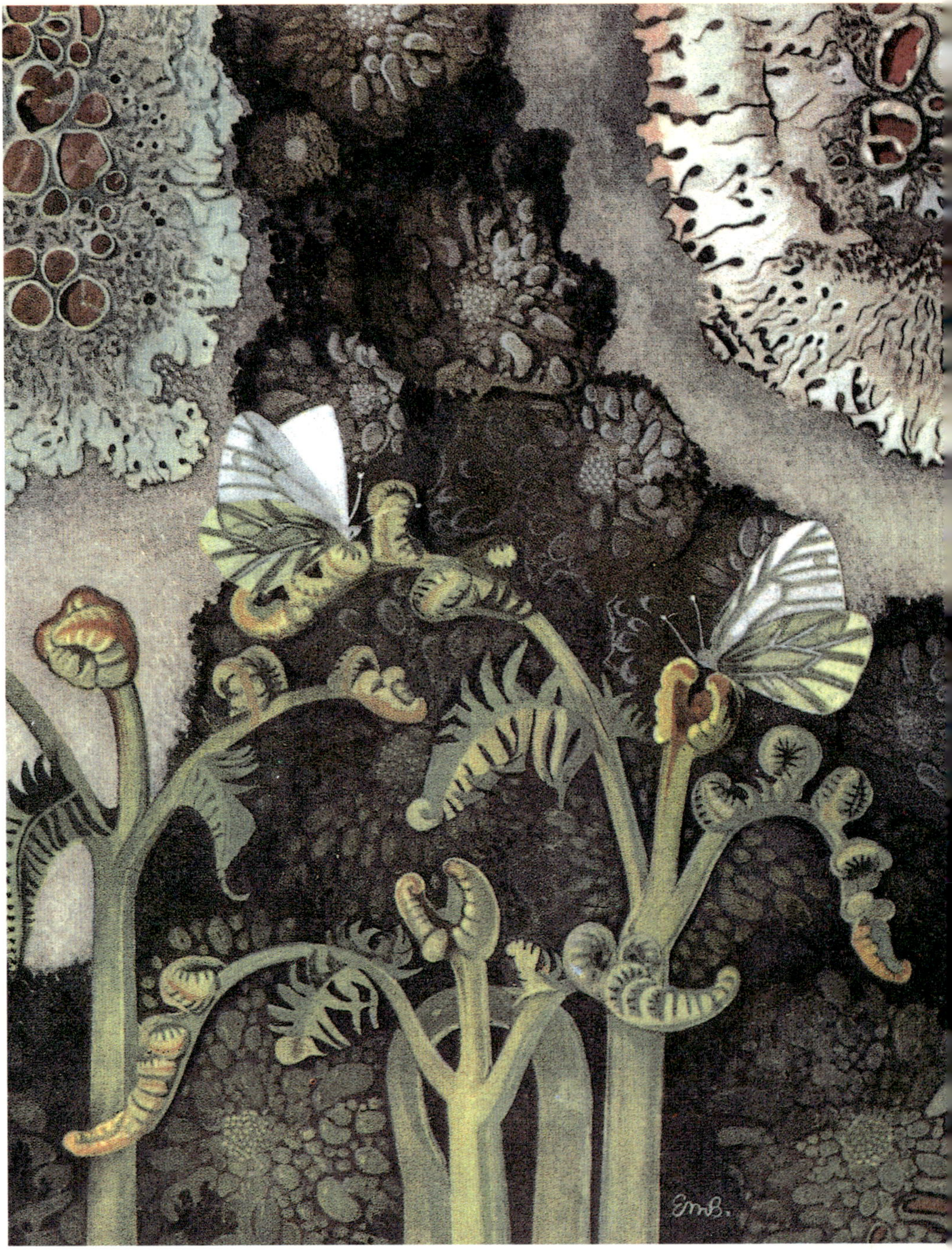
EmB.

PLATE V

*Parmelia conspersa**
(wet)

*Umbilicaria pustulata**
(dry)

Parmelia saxatilis
(fruited)

*Umbilicaria pustulata**
(wet)

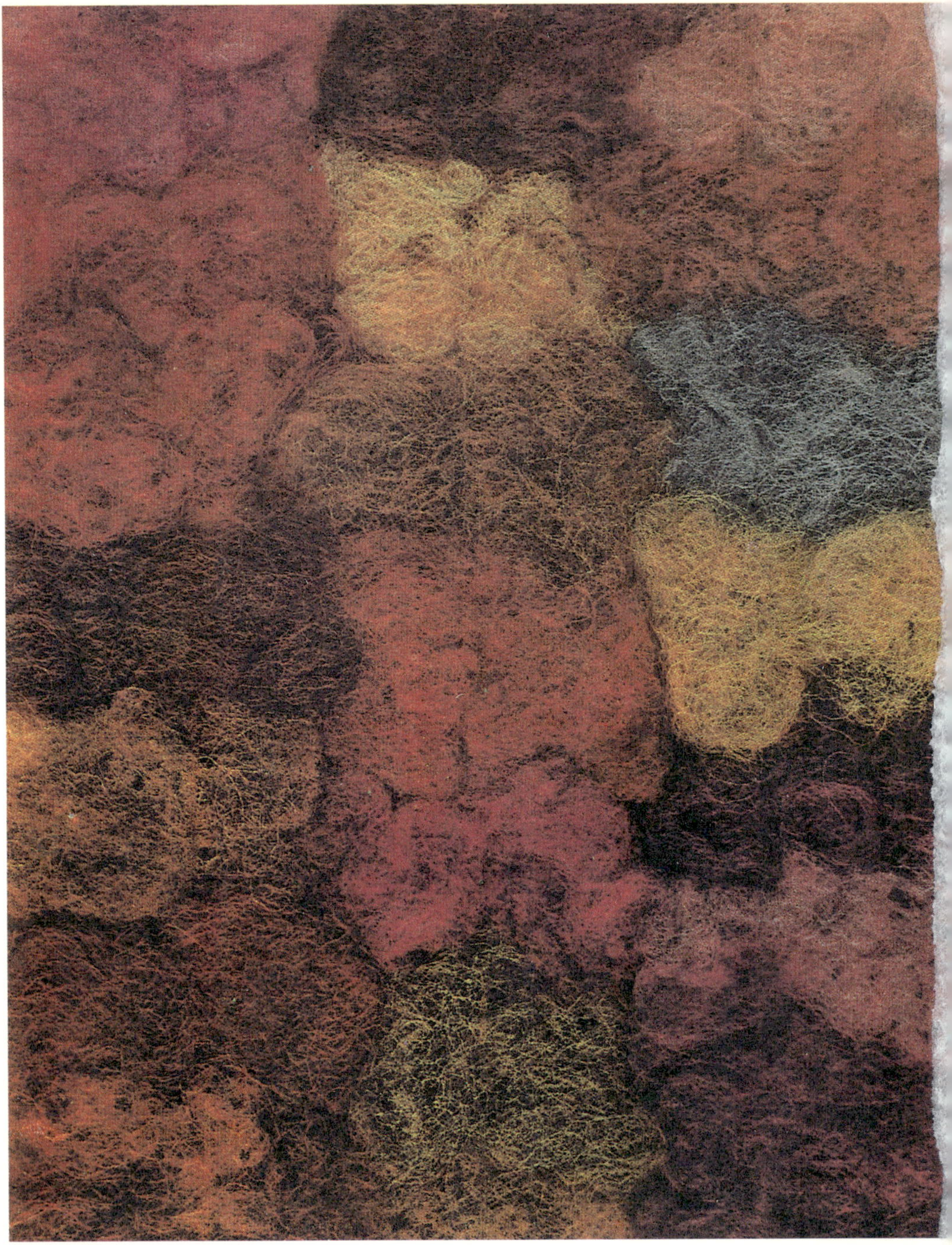

PLATE VI

Samples of Dyed Fleece

1 *O. tartarea* Orchil, soda	8 *E. prunastri* Orchil	15 *O. parella* Orchil, pale
2 *O. tartarea* Orchil	9 *E. prunastri* Boiling water	16 *X. parietina* Soda, ammonia
3 *O. tartarea* Orchil, alum	10 *L. pulmonaria* Boiling water	17 *X. parietina* Soda, ammonia (sun)
4 *P. omphalodes* Boiling water	11 *U. pustulata** Orchil	18 *P. caperata** Boiling water
5 *P. omphalodes* Boiling water	12 *U. pustulata** Orchil	19 *P. fuliginosa** Orchil, soda
6 *P. saxatilis* Boiling water	13 *U. pustulata** Orchil	20 *P. fuliginosa** Orchil, soda
7 *P. saxatilis* Boiling water	14 *H. physodes* Boiling water	21 *P. fuliginosa** Orchil, soda

*Asterisks following names indicate lichens that have been reclassified (renamed); consult List of Reclassified (Renamed) Lichens, pages 38-39.

CHAPTER V

EXTRACTING THE ORCHIL DYES

Orchil is obtained by fermenting lichen with ammonia, water and oxygen in a warm atmostphere. Use the back of boilers, stoves, radiators or airing-cupboards, where the maceration can be frequently stirred in passing. A greenhouse or low sunny window-seat are good places in the summertime. The temperature needs to be between 56-75 degrees Farhenheit throughout the fermentation period, which varies from three weeks to a full 28 days. Today, the use of household ammonia probably hastens Orchil maceration as compared with the old and primitive methods formerly used. The addition of chalk or lime towards the end of the fermentation period is necessary to give consistency to the mass if making it up into balls. In the Highlands these balls were wrapped in dock leaves and hung up in peat smoke, a piece being powdered when needed, and boiled up with a little alum. This form of dye will keep indefinitely, and is a dark bluish colour.

Orchil will dye wool, silk, wood, feathers and marble, and can be used on leather.

The form called Litmus is still used as a reagent, for Orchil reacts to alkalis and acids; Cudbear is now being used once more as an edible dye for foodstuffs.

Ochrolechia tartarea

It seems best to rub the dried lichen through a sieve or colander, when pieces of shale and other plants can be felt and removed, for any other lichen might spoil the dye. When this has been done, empty the powdered lichen into a wide and shallow bowl, making sure that you have a well-fitting lid or plate to cover it closely. Moisten the lichen thoroughly and then add by degrees an ammonia solution. One part of household ammonia to two parts water, made up beforehand, seems to work well, and the whole mass should be just stirrable. Seal the bowl, but examine the content shortly after as the mixture will have swollen, and may need more of the solution added to make it stirrable again. It should then be put in a warm place.

From now on until the colour begins to run the mass must be constantly stirred, five or six times a day, and the lid must always be replaced closely. The mixture should smell somewhat pungent, and if it has slackened, add a little more solution.

As soon as the colour begins to run and deepen, the mixture need only be stirred about three times a day. *O. tartarea* begins to run red like blackberry juice, and gradually takes on a dark bluish tinge, due to the ammonia. When the mixture is exposed to the air this blueness passes off slightly, and at the end of the fermentation period, which varies from 15 to 28 days, the lid is removed and the blueness driven off altogether.

If only the purples are wanted, then the fermentation can be stopped as soon as the colour has run full and strong. The dye can then be used straight away in the liquid state, or it can be made up into balls by adding chalk and allowing the mixture to dry out to a workable consistency. This is rather a messy, wasteful way of preserving Orchil, and the moth grubs love these balls. A better method is to allow the mixture to evaporate entirely in a warm place, when the residual powder can be stored in glass jars.

The addition of a little soda to the dye vat will turn the colour towards blue-purple, so that one can get a purple from the fully-matured red Orchil; but this is not so sharp as that from the maceration which has been stopped early on in the fermentation process. By the addition of Acetic Acid the purple dye can be made redder.

To dye with fully-matured Orchil, soft water should be used, and the addition of an acid should not be necessary if the *tartarea* has been well fermented.

When once the colour has begun to run well, it should not be necessary to add any more ammonia solution, and provided the bowl has a well-fitting lid or plate over it, the ammonia should not slacken.

In recipes of only 65 years ago, Cudbear and Orchil were still given, used either alone or with other vegetable dyes, on an Alum basis, with Glauber Salts. There are endless possibilities with this dye to bottom other brown lichen dyes, as well as other vegetable dyes.

One tablespoonful of this mixture, before it is dried, will dye about two ounces of wool.

Umbilicaria pustulata*

This lichen, being very hygrometric, will become tougher again in damp weather, even if it has been put away in sacks quite crisp and brittle. It is impossible to powder it in this state, and is best put in a very cool oven, or before a fire, until it is crisp once more. After this treatment it can be crumbled into smaller pieces by hand, and then be put through an ordinary mincer. Shake the mincer well afterwards, so as not to waste any of the powder, which should be put into a bowl.

The maceration can now proceed as already described for *O. tartarea*. A red may appear for a time around the rim of the bowl quite early on, but this will vanish again. The whole mass will next become a very dark brown, the red tinge not appearing for ten days or more.

If purples are wanted, the process can be stopped when the colour has run full and rich, or be left longer, and be finally aired off to obtain the red dye.

*U. pustulata** gives a deep, stronger red than *O. tartarea*. It dyes very evenly, and the exhaust bath will dye wool a brown-amethyst shade. There seem to be two colour elements in this dye, a red and a brown one; and for this reason it is probably better not to add more Orchil to the dye bath for subsequent dyeings, but to start afresh.

*Parmelia fuliginosa**

Dry the plants thoroughly and crumble up with the fingers. This is a pungent and strong Orchil, and the red colour may appear soon after the ammonia solution has been added.

Proceed in exactly the same way as for *O. tartarea*. If the dye is made up into balls it will need a little chalk to hold it together. This Orchil has a rather fruity aroma when it is simmered in the dye bath. It may be necessary to add acetic acid to get a red colour, while a pale violet can be obtained by the addition of a little soda to the dye-bath. Unlike most of the other ammonia dyes, this Orchil seems to be fast to sunlight. Tufts of dyed fleece left in strong sunshine for a couple of months have barely changed colour at all.

This Orchil was discovered by the writer while experimenting with *P. fuliginosa**. There must be many more of these lichens that are worth testing in this way.

Evernia prunastri

This lichen needs to be made crisp before it can be crumbled or powdered, and then it can be macerated in the same way as *O. tartarea*. It will yield a deep plum colour.

A pale yellow can also be got from this plant by the boiling-water method. The freshly gathered lichens can be shredded with kitchen scissors or dried plants crumbled.

Ochrolechia parella

This is not an easy plant to macerate. It seems to need a slightly stronger solution of ammonia; otherwise it becomes so unbearably fishy that one casts it away. The old recipe for making the French dye (see Bancroft) included arsenic and lime, but country people merely steeped it in stale urine to obtain a reddish or orange colour.

The dye may reside mainly in the fruits and as these are very hard, soak the plant in a little water and pound well before adding the ammonia solution. Once past the initial stage, it will give a dusky pink.

All the dyes mentioned in this book are substantive dyes, and need no mordant, but the colours can be varied by passing the dyed wool or silk through hot solutions of alum, chrome or acetic acid afterwards. A tin mordaunt may make Orchil more permanent, and change the colour more toward red.

Some Useful Hints

Ovenware glass dishes with lids are ideal both for macerating and for dyeing smaller quantities of wool or silk. Ammonia is a volatile alkali, and it is a help to keep the lid on when simmering a dye to which ammonia has been added.

Wool can be left in the dye to steep for a day or two if really deep shades are wanted. *P. fuliginosa** Orchil, to which a little soda has been added, will give a deep, royal purple if the wool or silk is left to steep for some time after dyeing.

Glass vessels will need a very low heat, an oil stove being best; but an asbestos mat on a low-temperature electric plate will prevent the glass from cracking.

Soda should always be used in small amounts, as wool never takes kindly to an alkali. If one wants to make the Orchil dye liquor more red, then the wool or silk must be lifted out while a little more acetic acid is entered. A tin mordant can be used in this way for making the colour faster.

If the mixture is stood in sunlight while macerating, the glass lids on ovenware act like burning glasses and help to keep the temperature up. Strong sunlight, however, is not good and a thin sheet of paper laid over the dishes helps to protect the mixture.

CHAPTER VI

EXTRACTING THE DYES, BOILING WATER METHOD

In Vegetable Dyeing, it is usual to simmer the leaves, twigs, roots, etc., in a muslin bag if they are going to remain with the wool or silk at a later stage. With lichens, this precaution is not necessary if the plants are kept in large pieces, as they can easily be shaken out afterwards. Lichens, however, take up a lot of room as they swell, so it is necessary to make allowance for their bulk, and to have containers big enough to take the wool, water and plants together.

Lichens yield far more dye if they are bruised well when moist, or given a tight grip in each hand to ensure cracking across when they are dry. The dye acids may reside in the upper cortex, between this and the algal layer; in the body of the plant; or in the lower cortex. The powdery and coral-like out-growths may also carry dye acids. For this reason it is better not to wash the plants first, as this valuable source of dye would be carried away in the waste water. Most lichens contain grit, insects, etc., but these will not affect the dye; moss and bark *must* be removed, however, as barks usually give a brown colour.

Galloe says that cells immediately below the upper cortex of *Parmelia caperata** are yellow. As these cells lie over the layer of green algal cells, they probably account for the very yellowish tint even when the plant is dry. *P. caperata* loves Alder trees, but it is a very tedious business to remove the flakes of bark, and it is better to gather this lichen from smooth-barked trees or stones. The usnic acid which the yellow cells contain gives a clean, clear dye, but an interesting experiment might be made by leaving the Alder bark on sometimes. A good yellow-brown may result.

Galvanized iron buckets, egg-preserving pails and two-handled washing baths are ideal for dyeing with these plants. Plain iron coppers with a fire beneath may tend to darken browns, and should not be used for yellow dyes.

The quantities usually given for this method are one pound of lichens to one pound of wool. This quantity applies to the plants in a dry state, and more should be allowed when the plants are moist, to off-set the water content. There are two traditional ways of dyeing with lichens.

The first method is to bring the lichens slowly up to the boil in soft water to which about one teaspoonful of Acetic Acid has been added for the above quality of lichens. Acid and peat-stained mountain water is even better. When boiling point is reached the heat is lowered and the whole left to simmer for three hours, after which the heat is turned off, and the dye left overnight to cool. The well-wetted wool is entered the next day, brought up to boiling point once more, and the whole simmered until the desired depth of colour has been obtained.

With the second method, the dye-bath is nearly three-quarters filled with alternate layers of first wool and then lichens. The bath is then filled with soft water made acid, and the mixture simmered until the desired shade is obtained. The process of bringing up to the boil and simmering can be done on two consecutive days if necessary, and the wool will not be harmed by leaving it in the bath, though allowance must be made for the possibility of the wool taking up more colour. With both these methods of dyeing, it is necessary to turn the wool over and about very gently, to ensure evenness in the colour. Fleece can be dyed by this method.

For small quantities of wool, dyed in oven-ware glass or similar small bowls, more colour can be got by shredding fresh and moist plants with kitchen scissors, or crumbling dried ones. Either the wool or the lichens may be kept apart in a very loose muslin bag. The plants are simmered until the colour runs, and the wet wool entered. When dyed, it can be left to cool. A little Acetic Acid will be needed to help extract the colour.

When the wool has been dyed and gone cold, it is washed in one or two waters and given a shaking to loosen the plants while they are still swollen and moist.

These lichen dyes do not need a mordant on wool, but linen will have to be mordanted first with alum if it is to be dyed yellow with *Peltigera canina*. This can be done with ¼ ounce alum to one pound linen, after the linen has been prepared in the usual way. Longer boiling will be necessary in both the mordanting and dyeing processes. White or natural carpet wool takes the brown lichen dyes very well, and a rug woven with this wool will always have an attractive aroma. A very beautiful red-brown can be got by using an Orchil dye over a dark *P. omphalodes* brown, or by dyeing a ground colour with Orchil first. Both *O. tartarea* and *U. pustulata** reds dye very evenly on carpet wool, and are lovely in themselves.

Lobaria pulmonaria should only be used for small quantities of wool, as it takes years to grow, and forestry operations are likely to make this plant and other lichens even more scarce. It might be possible to get permission to gather this plant from felled trees.

Xanthoria parietina will yield a golden-brown to boiling water, and various shades of pink or purplish-pink with an alkali. Caustic soda will excite the colour most, but this is so destructive of wool that great care has to be taken. On wool that has been first mordanted with chrome, a dusky pink can be obtained. When using ammonia, it is best to keep the dye-bath lidded, but the cover will have to be lifted a little if the dye threatens to boil over, and then replaced again.

Hypogymnia physodes and *Parmelia conspersa** are both traditional Crottles. *H. physodes* gives a yellow brown by boiling, and *P. conspersa** yields a strong orange-brown, rather similar to that from *P. saxatilis*.

Author's Experiment Producing a Blue from Xanthoria Parietina.

A two ounce tobacco tin of the dried plant was crumbled into an ovenware glass dish, capacity 9 fluid ounces. The dish was filled with soft water to which was added a good pinch of common soda. This was then set aside with the lid on for two or three days to steep. No pink colour appeared after this time so the mixture was set over an oil burner, brought very slowly to the boil, and then simmered a little longer, being kept lidded. Tufts of white fleece were then entered one at a time, and dyed a good pink colour in both dishes. The experiment was repeated on two more consecutive days, and as the quantity of dye was reduced by simmering, both lots were amalgamated into one dishfull.

It was noticed that the last tufts of pink fleece were turning a dull slate-blue in the strong sunshine outside. This was so puzzling that the dish was laid aside at the back of the stove, one small tuft of fleece being left in and the lid replaced. After about two weeks, the dye (smelling a little off), was carried outside into brilliant sunshine. On squeezing out the piece of fleece, and placing it on the window sill to dry, it turned a lovely clear blue the moment bright sunlight fell upon it. While more fleece was being sought, the sunlight passing through the glass lid warmed the small quantity of dye left, so that it steamed on raising the lid. It was now possible to enter bits of wool, leave them for several minutes and, on squeezing them out, to obtain the same clear blue. In fact, the colour was behaving rather like that from the Tyrian shellfish, except that the process was reversed, changing from reddish to blue, instead of from blueish to red.

Long boiling and long steeping afterwards, even to the point of the dye being a little fermented, seem to contribute to the success of this experiment, but obviously the main factor is a very brilliant sun. A piece of the blue wool was exposed to the same strong sunlight for a week, half being kept hidden. The colour lost a little brilliance, and then seemed to remain at a softer shade.

This experiment with *X. parietina* was again successful as far as the slate-blue stage, and then the sun refused to play its part. It would be better perhaps to leave the long-simmered dye a full 28 days before dipping and exposing to sunlight.

APPENDICES ON ACIDS

Much work is being done in North America on the acid-content of lichen plants. Mason E. Hale has proved that both *Umbilicaria pustulata** and *Ochrolechia frigida* contain the Orchil Acid, Gyrophoric Acid. When once it is known that these two plants give a similar dye, then any other lichens proved to contain the same acid are likely to yield a red dye. There may be variations, of course, due to the presence of other substances. Hale says that *U. pustulata** also contains an unknown red substance. This may account for the different reaction of *U. pustulata** to the process of fermentation, noticed in the experiments given in this book. The mixture became a very dark brown at one stage, and the exhaust-bath dyed wool a brown-amethyst shade. This plant was used at one time in making a black paint.

Having proved in this book that *Parmelia fuliginosa** can be made to yield a good Orchil, it was interesting to learn that this plant contains another Orchil acid, Lecanoric Acid. As this Orchil appears to be faster to sunlight than Gyrophoric Orchil, it will be of great help to handweavers to know of other plants containing this acid. It might be good idea to combine the two Orchils in the dyebath. Below is a list of lichens found in Britain and America, which have been proved by Hale to contain these two acids.

Parmelia saxatilis and *P. omphalodes* contain two acids, Atranorine and Salacinic Acids, but *P. omphalodes* contains one other, Lobaric Acid. *P. omphalodes* always dyes a darker and faster brown than *P. saxatilis*. In Iceland this lichen has been made to yield a pale Orchil, by steeping it in a Potash solution, and adding salt.

Lichens containing Usnic Acid are likely to give good clear yellows, but this colour may be altered by the presence of other acids in the plants. For instance, *Hypogymnia physodes* contains Usnic Acid, but it also contains three other acids and so yields a yellow-brown to boiling water.

Ammonia fermentation may extract one acid colour, and boiling water another acid colour from the same plant. Thus fermented, *Evernia prunastri* will yield an Orchil from the Evernic Acid which it contains, and a pale yellow from another acid, by the boiling-water method.

Some lichens have chemical strains which means that a lichen may have different acids in each geographical area or locality. In North America, *Umbilicaria papulosa** may be a chemical strain of *U. pustulata** which occurs in the south and west of England, in Wales and in Scotland. Both of these lichens contain Gyrophoric Acid. A much more rare but similar lichen in North America is *U. pensylvanica*.

LICHENS COMMON TO BOTH COUNTRIES, KNOWN TO CONTAIN ORCHIL ACIDS IN AMERICA

Those marked with a double asterisk are chemical strains, and may of course present another Acid in Britain.

EVERNIC ACID *Evernia prunastri.*

LECANORIC ACID
All *Diploschistes spp., D. scruposa** (yields Orchil). *Parmelia dubia*, P. fuliginosa** (proved to yield Orchil). *P. furfuracea* *** (Lindsey said this plant gave an Orchil, but it may not have been a British specimen). *Parmelia laevigata* **, P. olivacea (P. olivetorum** gives a red reaction in Britain). *P. scortea*, P. subaurifera*.*

GYROPHORIC ACID
All *Gyrophora spp*.,* excepting *G. cylindrica.* (This is also stated by Lindsey. He says *G. cylindrica* gives a greenish-brown dye, probably with boiling water only). *Umbilicaria pustulata**, sometimes known as *U. papulosa** in America (proved to yield Orchil). *Ochrolechia androgyna, O. tartarea* both yield Orchil.

NOTE: There is the possibility that *Evernia furfuracea** might contain OLIVETORIC ACID in some areas in Britain, but it may only present its third acid, PHYSODIC, which does not give Orchil.

LICHENS CONTAINING THE ORCHIL ACIDS IN NORTH AMERICA

1 LECANORIC ACID

*Diploschistes scruposa**
*Lecidia friesii**
*Lecidia scalaris**
*Parmelia andreana**
*Parmelia bolliana**
*Parmelia cladonia**
*Parmelia dubia**
*Parmelia fuliginosa**
Parmelia furfuracea
*Parmelia laevigata**

*Parmelia manshurica**
*Parmelia olivacea**
*Parmelia rudecta**
*Parmelia subaurifera**
*Parmelia tinctorum**
Pertusaria bryophaga

2 GYROPHORIC ACID

Cetraria delisei
Dactylina arctica
*Lecanora gelida**
*Lecidia pelobotryon**
Lecidia tenebrosa
Lobaria erosa
Lobaria querzicans

Ochrolechia androgyna
*Ochrolechia frigida**
Ochrolechia yasudae
*Parmelia bolliana**
*Parmelia pseudoborreri**
Rhizocarpon grande
*Rinodina oriena**

NOTE: *Parmelia bolliana** produces both acids. Other dye-producing acids include EVERNIC ACID, contained in *Evernia prunastri*, and ERYTHRIN, which is in *Roccella* and *Diploschistes spp.* OLIVETORIC ACID occurs in *Parmelia cetrariodes**, *Cornicularia divergens**, *Cetraria ciliaris** and *Parmelia lobulifera*.

*Parmelia manshurica** and *Dactylina arctica* contain USNIC ACID and should give yellow by boiling, unlike the others with an ORCHIL ACID content.

ACID CONTENT OF THE BOILING WATER LICHENS

Lichen acids are known as "substances", and it is from these depsides and depsidones that dyes are actually made. Most lichen acids are colourless, aromatic compounds; however, there are two which impart a yellowish tinge to lichens: parietin and vulpinic acid.

The following list of lichens shows which acids different species contain. This information was provided by Mason E. Hale, from his *Lichen Handbook* published by the Smithsonian Institution in Washington, D.C.

ATRANORIN: *Evernia prunastri, Parmelia omphalodes, Parmelia saxatilis*

ERYTHRIN: various species of *Roccella* and *Diploschistes*

FUMARPROTOCETRARIC: various species of *Bryoria* and *Cladonia*; *Parmelia caperata**

GYROPHORIC: all species of *Umbilicaria* except *U. cylindrica*

LECANORIC: *Parmelia rudecta*, Parmelia subaurifera** and many species of *Lecanora*

NORSTICTIC: *Lobaria pulmonaria*

SALAZINIC: *Parmelia conspersa*, Parmelia omphalodes, Parmelia saxatilis* and some species of *Usnea*

USNIC: some species of *Alectoria* and *Ramalina; Parmelia caperata*, Parmelia conspersa**

VULPINIC: *Letharia vulpina*

Editors' Note: See bibliography for additional references containing information on lichen acids.

LIST OF RECLASSIFIED OR RENAMED LICHENS

Editors' Note: Where nomenclature differs in the British Isles and North America, synonyms (based on Hawksworth, James and Coppins for United Kingdom; Egan for United States) are given.

Cetraria ciliaris Ach. = *Tuckermannopsis ciliaris* (Ach.) Gyelnik (Lai 1980)

Cornicularia divergens Ach. = *Bryocaulon divergens* (Ach.) Karnef.

Diploschistes scruposa = *D. scruposus* (Schreber) Norman = *Urceolaria scroposa* (Schreber) Ach.

Gyrophora spp. = *Lasallia spp.* or *Umbilicaria spp.*

Lecanora calcarea (L.) Sommerf. = *Aspicilia calcarea* (L.) Mudd.

Lecidia friesii Ach. in Liljeblad = *Hypocenomyce friesii* (Ach. in Liljeblad) P. James and G. Schneider in Schneider

Lecidia pelobotyron (Wahlenb. in Ach.) Leighton = *Amygdalaria pelobotyron* (Wahlenb. in Ach.) Norman

Lecidia scalaris Ach. ex Liljeblad = *Hypocenomyce scalaris* (Ach. ex Liljeblad) M. Choisy

Lecidia tenebrosa Flotow = *Schaereria tenebrosa (Flotow)* Hertel and Poelt (Hawksworth et al. 1980)

Lobaria erosa (Eschw.) Nyl. = *L. ravenelli* (Tuck.) Yoshim.

Ochrolechia pallescens (L.) Massl. = *O. pseudotartare* (Vainio) Vers. (fide Tucker)

Parmelia andreana Müll. Arg. = *Flavopunctelia flaventior* (Stirton) Hale.

Parmelia bolliana (Müll) Arg. = *Punctelia bolliana* (Müll. Arg.) Krog

Parmelia caperata (L.) Ach. = *Pseudoparmelia caperata* (L.) Hale = *Flavopunctelia caperata* (L.) Hale

Parmelia cetrariodea (Delise ex Duby) Nyl. (Santesson 1984) = *Cetrelia cetrariodes* (Delise ex Duby) Culb. & C. Culb.

Parmelia cladonia (Tuck.) Du Reitz = *Pseudoevernia cladonia* (Tuck.) Hale & Culb.

Parmelia conspersa (Ach.) Ach. = *Xanthoparmelia conspersa* (Ach.) Hale

Parmelia dubia (Wulfen in Jacq.) Schaerer (Santesson 1984) = *Punctelia subrudecta* (Nyl.) Krog

Parmelia fuliginosa subsp. *fuliginosa* (Fr. ex. Duby) Laundon = *Melanellia fuliginosa* (Fr. ex Duby) Essl.

Parmelia furfuracea (L.) Ach. = *Pseudoevernia intensa* (Nyl.) Hale & Culb.

Parmelia laetivirens = *Parmelia fuliginosa* var. *laetivirens* (Flotow ex Korber) Nyl.

Parmelia laevigata (Sm.) Ach. = *Hypotrachyna laevigata* (Sm.) Hale

Parmelia lobulifera Degel. = *Hypotrachyna imbricatula* (Zahlbr.) Hale

Parmelia manshurica Asah. = *Flavopunctelia soredica* (Nyl.) Hale

Parmelia olivacea (L.) Ach. = *Melanelia olivacea* (L.) Essl.

Parmelia olivetorum (Nyl.) = *Cetrelia olivetorum* (Nyl.) Culb & C. Culb.

Parmelia pseudoborreri Asah. = *Punctelia borreri* (Sm.) Krog

Parmelia rudecta Ach. = *Punctelia rudecta* (Ach.) Krog

Parmelia scortea (Nyl.) = *Bulbothrix goebelii* (Zenker) Hale

Parmelia subaurifera (Nyl.) = *Melanelia subaurifera* (Nyl.) Essl.

Parmelia tinctorum Delise ex Nyl. = *Parmotrema tinctorum* (Delise ex Nyl.) Hale

Platysma glaucum (L.) Nyl. = *Platismatia glauca* (L.) Culb & C. Culb.

Platysma juniperina = *Cetraria juniperina* (L.) Ach.

Rinodina oriena (Ach. Massal. = *Dimelaena oriena* (Ach.) Norman

Umbilicaria pustulata (L.) Hoffm. = *Lasallia pustula* (L.) Mérat

Umbilicaria pustulata (L.) Hoffm. var. *papulosa* (Ach.) Tuck = *Lasallia papulosa* (Ach.) Llano

Urceolaria spp. = *Diploschistes spp, Pertusaria spp.*

BIBLIOGRAPHY

Bancroft, Edward. *The Philosophy of Permanent Colours*. First edition. London. 1794.

Dahl, Eilif. *Analytical Keys to British Macrolichens*. Cambridge, 1952.

Edge, Alfred. 'Some British Dye Lichens.' *Journal of the Society of Dyers and Colourists*, May, 1914.

'Archils, Dyes, Lichens'. *Encyclopedia Britannica*. 1798.

Galloe, O. *Natural History of the Danish Lichens*. Copenhagen. 1927-1972. 10 vols.

Lindsey, W. Lauder. 'Tinctorial Properties of Lichens,' *Edinburgh Philosophical Journal (New)*. No. 42. 1854.

Lindsey, W. Lauder. *British Lichens*.

Mairet, Ethel. *Vegetable Dyes*. London. 1952.

Smith, Annie Lorraine. *Monograph of the British Lichens*. London. 1911. 2 vols.

Smith, Annie Lorraine. *Handbook of British Lichens*. London. 1921.

EDITORS' BIBLIOGRAPHY

Brodo, Irwin M. *Lichens of the Ottawa Region*. Ottawa. 1988.

Egan, Robert S. *Fifth Checklist of the Lichen-forming, Lichenicolous and Allied Fungi of the Continental United States and Canada*. American Bryological and Lichenological Society, Inc. 1987.

Hale, Mason E. *How to Know the Lichens*. Second edition. Dubuque, Iowa. 1979.

Hawksworth, D.L., P.W. James and B.J. Coppins. *Checklist of British Lichenforming, Lichenicolous and Allied Fungi*. British Lichen Society. 1980.

Jahns, Hans Martin. *Ferns, Mosses and Lichens of Britain, North and Central Europe*. London, 1983.

Llano, George Albert. *Monograph of the Lichen Family Umbilicariacae in the Western Hemisphere*. Washington. 1950.

Richardson, D.H.S. *The Vanishing Lichens*. Newton Abbot. 1975.

Seaward, M.R.D. *Provisional Atlas of the Lichens of the British Isles*. Bradford. 1985.

Seaward, M.R.D. and C.J.B. Hitch. *Atlas of the Lichens of the British isles*. Cambridge. 1983.

Thomson, John W. *American Arctic Lichens. Vol. 1. The Macrolichens*. Irvington-on-Hudson. 1984.

INDEX

Acid content of the Boiling-water Lichens	32,35-38
American Orchil Lichens	33-37,*passim*
Ammonia, sources of, in solution	10,17-18,25,28-31,34
Blue dyes	5,10,25,34
Boiling-water Lichens	10,19-20,26,28-29
Hypogymnia physodes	19,34,35,Plates I,III,VI
Lobaria pulmonaria	19-20,33,38,Plates I,III,VI
Parmelia caperata*	19,21,32,38,40,Plates I,III,VI
Parmelia conspersa*	19,21,32,38,40,Plates I,V
Parmelia omphalodes	10-11,14,19,23,33,35,38,I,II,IV,VI
Parmelia saxatilis	10-11,13,19,22-23,35,38,Plates I,II,VI,V
Peltigera canina	19,24,33,Plates I,II,III
Platysma glaucum*	19,41,Plate IV
Xanthoria parietina	13,19,23,26,33-34,Plate VI
Boiling-water methods	8,32-34
Brown dyes	6-8,10-11,18,20-23,26-27,29-30,32-35
Extracting the Orchil dyes	8,17,28-31,35
Gathering and drying plants	14,17
Lichen plant, spores, etc.,	12-13,Plate I
Less plentiful dye Lichens	27
Orchil, Litmus, Cudbear, History of,	9,16-17,28-29
Orchil-producing Lichens	7-8,15-17,35,Plate III
Evernia prunastri	15,30,35,37-38,Plates I,III,VI
Ochrolechia parella	10,15-16,30,Plates I,II,VI
Ochrolechia tartarea	9-10,13-14,15-17,29-30,33, I,IV,VI
Parmelia fuliginosa*	13,15,17,30-31,35,37,40,Plates IV,VI
Umbilicaria pustulata*	9,12,15,18,29-31,33,35,38,41, I,II,V,VI
Lichens common to Britain and North America	14,36,*passim*
Purple, red and violet dyes	5-10,13,15-18.23-27,29-31,33-34,36
Rare and endangered lichens	14
Useful hints	31
Wool, fleece and silk dyeing	31,32-34,*passim*
Yellow dyes	8,10,11,13,20,21,24-27,30,32-35,38

Other titles published and distributed by Robin & Russ Handweavers are

1. NATURAL DYES FROM NORTHWEST PLANTS - *Judy Green* - *10.00-P*

A beautifully organized paperback of clear, brief, practical recipes for outstanding dyes from about 50 weeds, trees, and seven flowersd which grow wild across North America. Basic instructions are given for beginners covering equipment, fiber preparation, mordants, collecting dyestuffs, preparing dyebaths, and some "Trade Secrets" that will interest advanced dyers as well.

2. HANDSPINNER'S WORKBOOK: FANCY YARNS - *Mabel Ross* - *19.95-P*

Spinning and plying simple yarn is enjoyable and therapeautic. If you are ready to create more exciting yarns, you can by using this book from the author of the popular books: "The Essentials of Handspinning" and "The Essentials of Yarn Design". This workbook introduces methods of blending colors and spinning more exotic fibers such as camel, alpaca, angora, mohair, silk, flax, and more; Learning special twists to make pigtails, loops, snarls, ric-racs, marls, and thicks-n-thins.

3. SPINNING WHEELS, SPINNERS & SPINNING - *Patricia Baines* - *12.95-P*

Now in its 3rd edition, this is a thoroughly researched study of spinning history and folklore since the 13th century considered by many to be one of the best spinning books. Both instructional and educational, this book also contains practical step-by-step instructions.

4. CARDWEAVING OR TABLET WEAVING - *Russell Groff* - *7.00-P*

Now in its seventh printing, this is the most complete cardweaving instructional book available. It begins with concise diagrams showing how the cards are used, followed by brief explanations of pattern drafts, set-up, and weaving. Then the book is devoted to giving 53 creative patterns: each pattern accompanied by a photo and a diagram. A final chapter shows you how to draw-down a pattern to reproduce an existing piece.

5. CRACKLE WEAVE, THE - *Mary Snyder-9.50-P*

Now with 90 PATTERNS! Complete drafts for 4,6,8, and more harness patterns. Her block arrangements create some of the most facinating and unusual patterns. Those who have her first edition will be inspired by the 40 NEW PATTERNS in this edition. Mary Snyder is also the author of our popular book: ***Lace & lacey Weaves.***

6. ESSAY UPON THE SILKWORM - *Henry Barham* - *12.95-H*

We reprinted this book from its 1719 original. King George had this published to stimulate silk production in the British Empire to compete with the orient's silk monopoly. It explains in great detail how to raise silkworm, feed them, and produce finished silk as it was done in the 18th century. You will be amazed how herbs and spices were used effectively as insecticides.

7. EYE FOR COLOUR - *Bernat Klein* - *20.00-H*

A creative approach using his original color thories in which fabric designs are inspired from paintings and other mediums. Featured in full color photos are paintings of Seurat, Monet, and Klee. Bernat Klein, a world famous designer of fabrics and yarns, explains how aspects of his life have influenced his designs. He also reveals his well-received theory that an individual's entire wardrobe should be based in their eye color and the eye's role in your appearance.

8. INKLE WEAVING - BRADLEY - *15.95-P*

A comprehensive guide to explain the inkle loom's unrealized potential. It also shows weavers with large looms how an inkle loom can be a compliment to their workshop. Clearly describes how to use an inkle loom including detailed plans to make your own inkle loom.

9. JAPANESE IKAT WEAVING - *Tomita - 24.95-P*

Also known as Kasuri, Japanese Ikat weaving is a technique in which lengths of yarn are tied and dyed before weaving. This book gives a history of Kasuri and teaches you the various Ikat techniques developed in different regions of Japan.

10. LACE AND LACEY WEAVES - *Mary Snyder - 8.50-P*

This revised edition contains 86 projects with complete worksheets for all including both four and multiple harness projects. You will learn to make pick-up lace, Bronson, Myggtail, Swedish Lace, and barley corn weaves.

11. DRESSING THE LOOM - *Ida Grae - 6.00-P*

An award winning book! If you're looking for instructions for warping that is easy to understand, this is it. The direct beaming method of warping is explained in detail with numerous step-by-step photographs.

12. 1000 + PATTERN IN 4, 6, & 8 HARNESS SHADOW WEAVES - *Marian Powell -12.95-P*

Tremendous research went into this book containing 650 4-harness, 300 6-harness, & 300 8-harness patterns; each pattern has weaving instructions accompanied by a photo. We have woven several of these patterns for our monthly bulletins receiving compliments for each of them.

13. 16 HARNESS PATTERNS - *The fanciest twills of all - Irene Wood - 10.00-P*

An intense study of Fred Pennington's weaves with detailed photos and all the instructions needed to reproduce 150 eye-catching 16 harness patterns. Each pattern has a full-page photo of the resulting fabric. Included is a section showing you how to create your own 16 harness patterns.

14. TARTAN WEAVER'S GUIDE - *James Scarlett - 9.95-H*

228 different tartan thread counts are provided along with FULL COLOR PHOTOS OF 142 COMPLETED TARTAN FABRICS. The author is a true Scotsman and a handweaver specializing in tartans. During a visit to Scotland, he came to the airport to greet us weaving his tartan skirt. Very few publications give the thread counts as accurate as this book.

15. SECTIONAL WARPING MADE EASY - *Russell E. Groff - 7.00-P*

Frustrated with tangled warps and afraid to do long warps? Let Mr.Groff show you how to avoid these problems with sectional warping. This manual uses many photographs to show the step-by-step stages of sectional warping. Text and close-up photos explain concisely this time-saving process. Excellent for beginners.

16. 200 PATTERNS FOR MULTIPLE HARNESS LOOMS - *Russell Groff* - *12.95-P*

This book compiles the most popular fabric designs that appeared in the first 25 years of ***Drafts and Designs***, "Our multiple harness weaving bulletin". In this book are examples of an incredible variety of 5 to 12 harness pattern techniques. Each pattern includes complete directions of weaving, threading, treadling, tie-up, and threads used with a photo of each resulting fabric.

17. THE ESSENTIALS OF HANDSPINNING - *Mabel Ross* - *7.95-P*

This well-written introductory booklet was printed in an inexpensive form to help you decide if you would enjoy handspinning without much initial investment. It has complete information about selecting fleeces, wool preparation, using a handspindle, selecting a spinning wheel, and the types of yarns that can be made with different styles of spinning.

18. THE ESSENTIALS OF YARN DESIGN - *Mabel Ross* - *14.95-P*

A clearly written, practical handbook giving the details needed for planning and spinning exactly the yarn you want for a specific project. Describes in detail how to prodcue all kinds of expensive FASHION YARNS.

19. BASIC DOUBLE WEAVE THEORY - *Sara Farrar* - *8.00-P*

This booklet contains clearly written, step-by-step instructions for reproducing 4, 6, and 8-harness double weaves. Detailed diagrams show each step for creating 2, 3, or 4 Layered fabrics; Double, Triple, and Quadruple Widths; Fancy Tubes, and Pockets. It is spiral bound as an easy-access workbook.

20. MORE THAN FOUR: *A Guide for Multiple Harness Weavers* - *Mary-Elizabeth Laughlin* - *12.95-P*

Highly recommended by Peter Colingwood, an outstanding book for serious multiple harness weavers. Many different classes of four-harness patterns are expanded to eight-harness or more. Pattern designing rules and limitations of weaving structures are carefully explained including: multi-layered fabrics, wadded double-cloth, Bedford cord, pique, and center stitched doublecloths. You won't find many patterns to copy here. This book provides a sturdy foundation for creating and understanding woven fabric structures.

21. IBAN OR SEA DAYAK FABRICS & THEIR PATTERNS - *Hadden/Start* - *14.95 -P*

Iban of Sarawak textile techniques are studied in detail presenting the Cambridge Museum of Archaeology and Anthropology Iban Collection. This is a reprint of 1936 monograph re-illustrated with new photos and diagrams.

22. STUDIES IN PRIMITIVE LOOMS - *H. Ling Roth* - *7.95-P*

Reprinted from original Bankfield Museum Notes, this edition is much clearer than the original version. With new photos and drawings, the African looms are well-illustrated and their functions explained as well as looms from the Solomon Islands, Indonesia, and many more! This book also includes photos featuring amazing fabrics that were created without modern technologies.

23. THE WARP WEIGHTED LOOM - *Marta Hoffmann* - *16.95-P*

Tremendous research went into this book revealing the development and spread of warp-weighted looms through Europe beginning in Norway during the Dark Ages. The author also includes instructions using clear diagrams and photos to answer many questions about one of the oldest known weaving methods.